Do Brilliantly

AS Physics

Mike Bowen-Jones

Series Editor: Jayne de Courcy

Published by HarperCollins*Publishers* Limited
77–85 Fulham Palace Road
London W6 8JB

www.**Collins**Education.com
On-line support for schools and colleges

First published 2001

ISBN 0 00 710706 4

British Library Cataloguing in Publication Data
A catalogue record for this book is available from the British Library

Edited by Kathryn Senior
Production by Kathryn Botterill
Cover design by Susi Martin-Taylor
Book design by Gecko Limited
Printed and bound by Scotprint

Acknowledgements
The Author and Publishers are grateful to the following for permission to reproduce copyright material:
Oxford, Cambridge and RSA Examinations (OCR) (Chapters 1, 2, 3, 5, 6, 7, 10). Answers to questions taken from past examination papers are entirely the responsibility of the author and have neither been provided nor approved by the Oxford, Cambridge and RSA Examinations.
AQA (Chapters 1, 2, 5, 6, 7, 8, 9, 12, 13, 14, 15). AQA accepts no responsibility whatsoever for the accuracy or marking of the answers given.
Edexel Foundation, London Examinations (Chapter 1, 3, 4, 8, 10, 11, 13, 15). Edexel Foundation, London Examinations, accepts no responsibility whatsoever for the accuracy or marking of the answers given.
Welsh Joint Education Committee (WJEC) (Chapters 9, 11, 12). Answers to questions taken from past examination papers are entirely the responsibility of the author and have neither been provided nor approved by the Welsh Joint Education Committee.

Illustrations
Cartoon Artwork – Roger Penwill
DTP Artwork – Richard Morris

Every effort has been made to contact the holders of copyright material, but if any have been inadvertently overlooked, the Publishers will be pleased to make the necessary arrangements at the first opportunity.

Contents

How this book will help you

by Mike Bowen-Jones

Exam practice – how to answer questions better

This book will help you to improve your performance in your AS Physics exam.

Every year I mark exam papers where students don't use the information that they've learnt as effectively as they could. This means they don't get the grade they're capable of achieving.

To get a high grade in AS Physics you need a good grasp of the subject matter and good exam technique. Your textbook can help you develop your knowledge and understanding. **This book will help you improve your exam technique, so that you can make the most effective use of what you know.**

Each chapter in this book is broken down into four separate elements, aimed at giving you the guidance and practice you need to improve your exam technique:

1 Exam question, Student's Answer and 'How to score full marks'

Each chapter begins with a question and a typical student's answer. Many of the students' answers are quite good. But they all have the sorts of mistakes that are common and which you can avoid fairly easily (provided you do understand the topic quite well!). **The 'How to score full marks' section shows how to correct the mistakes that the student has made and how to pick up extra marks.**

2 'Don't make these mistakes'

In these boxes I highlight many of the most common mistakes that I see each year in students' exam scripts. Wherever possible, I have related these to the student's answer that you have just gone through. However, these ideas also stand alone. **You may want to read through these boxes just before sitting your exam to make sure you avoid these mistakes when answering questions**.

3 'Key points to remember'

These key points are not intended to replace your notes or textbook. I've included them **as a quick reference guide to some of the most important facts you really MUST know for your exam**. You could also read these through as last-minute revision to help bring the main ideas to the front of your mind just before sitting your exam.

4 Questions to try, Answers and Examiner's hints

Each chapter has at least one exam question for you to try answering. Try doing these 'cold' (without using the hints) first – just like in the exam. This will give you a good idea of how much you really do know. Be honest with yourself!

If you need a little guidance then refer to the hints. These should put you on the right track. When you've completed your answers, look at the answers at the back of the book, together with my comments. These answers would gain full marks but they may not be the only ones that would do so and they may not be exactly the same as your own. You may need to consult with teachers, tutors or other people on your course to help you to decide whether or not what you have written is the equivalent of the answer provided.

The topics covered by your AS specification

The exam questions in this book are selected from specimen AS exam questions provided by the four exam boards AQA, Edexcel, OCR and WJEC.

Although each of the exam boards must include a 'common 60%' within each of their specifications, there are some either/or options even within that 60%! This means that different exam boards set questions on different topics. **I have attempted to provide questions on all the main 'common core' topics**. The grid below shows you which of the topics in this book are relevant to your Board's AS specification.

	AQA – A	AQA – B	EDEXCEL	EDEXCEL SALTERS HORNERS	OCR – A	OCR – B	WJEC
Chapter 1	✓	✓	✓	✓	✓	✓	✓
Chapter 2	✓	✓	✓	✓	✓	✓	✓
Chapter 3	✓	✓	✓	✓	✓	✓	✓
Chapter 4	✓	conservation of energy	✓	work, energy and power	work, energy and power	work, energy and power	●
Chapter 5	Young's modulus	●	*	✓	✓	✓	✓
Chapter 6	✓	✓	✓	✓	✓	✓	✓
Chapter 7	✓	✓	✓	✓	✓	✓	✓
Chapter 8	✓	✓	✓	✓	✓	✓	✓
Chapter 9	●	✓	●	✓	✓	✓	✓
Chapter 10	●	✓	●	✓	✓	✓	✓
Chapter 11	refractive index	✓	●	✓	✓	✓	✓
Chapter 12	✓	spectra	●	✓	photoelectric effect	✓	✓
Chapter 13	isotopes α-scattering	✓	✓	✓	●	●	nuclear structure
Chapter 14	✓	✓	*	●	●	●	●
Chapter 15	molecular kinetic theory	●	✓	specific heat capacity	●	●	●

✓ This topic is largely covered in this specification
● This topic is largely not covered in this specification
* This topic is available as an option in this specification

- Many of the exam scripts I mark contain errors that could be avoided if the candidate followed a few relatively straightforward ideas. **Examiners would like you to develop is a set of 'good habits' that you apply as 'second nature'**. You can then concentrate on making the most of the physics that you know.
- Questions offering more marks require **more detailed answers** than those worth fewer marks.
- Every numerical answer needs **an appropriate unit** (unless the quantity is 'dimensionless').
- You must communicate your thoughts to the examiner **in a clear way**.
- Don't panic and rush through the exam without taking the time to **think through your answers**.
- **Make sure your handwriting is legible**. The examiner will try hard to read what you have written, but if he or she cannot then you will not gain any marks. Don't forget that, unlike your teacher or tutor, this is the first time the examiner will have seen your handwriting and so you should go all out to make his or her task as easy as possible.
- Take time to **read the set of instructions on the front of the exam paper** that tells you how to approach answering questions. You must follow these instructions precisely.
- **Make sure you understand the meaning of the words used in exam questions that 'command' you to respond in a particular way**. If you don't, you are likely to answer questions incorrectly and miss out on a significant number of marks.
  - **State** – means that you need to recall a name, a phrase or an equation. It needs no explanation and the answer is usually worth one mark.
  - **Define** – means that you must recall and write down a formal 'textbook-type' of statement. This is likely to be worth either one or two marks.
  - **Explain** – requires some detail. You need to give a concise but relevant statement of the meaning of a concept in a manner that is clear and unconfusing. Don't waffle! In physics, it is quite acceptable to 'explain' things mathematically using standard mathematical symbols.
  - **State and explain** – Questions will often ask you to state and explain something. The 'state' part will probably be worth a mark or two and the 'explain' part the remainder of the marks.
  - **Discuss** – usually requires you to consider different aspects of a situation. This type of question is very open-ended and difficult to mark consistently since there is no single 'correct' answer. Questions with this command word are likely to be worth a significant number of marks.
  - **Describe** – usually relates to experimental techniques and needs no more than a simple description of what you would do, without any supporting reason. Diagrams or sketch graphs will often help to focus your description. Questions with this command word may also be worth a high number of marks.
  - **Sketch** – is a term used almost uniquely in physics and requires you to draw the general shape of a graph. You must label the axes with quantities and units (if possible). Be careful to include the origin if it is relevant and any other key points should also be shown. Questions like this may also carry quite high marks.
  - **Show that** – means that you are given the answer and need to provide the full argument (usually mathematical) that shows why the given answer is correct. You must include a full description of how you reached the answer if you are going to score full marks. These questions are usually worth two or three marks.
  - **Suggest** – means that you may not have studied this topic but should be able to come up with a sensible reason or answer based on what you know. There is unlikely to be only one correct answer. These questions may be worth anything between two and a lot of marks. You can tell from the marks available, how detailed your answer needs to be.
  - **Estimate** – means that you are not expected to work to more than one or two significant figures using either rough values provided in the data or else values that you need to make a reasoned 'guess' about. Often these questions require answers that are no more than the correct order of magnitude (power of 10). 'Estimate' calculations are not detailed and so are usually worth fewer marks than 'normal' calculations.

Exam question and student's answer

1 (a) (i) Define acceleration.

v/t ∧ ∧ — 0/2

(ii) State the unit for acceleration.

ms^{-2} ✓ — 1/1

[3 marks]

(b) Fig. 1.1 shows the variation with time t of velocity v for a short journey travelled by a car of mass 800 kg.

Fig. 1.1

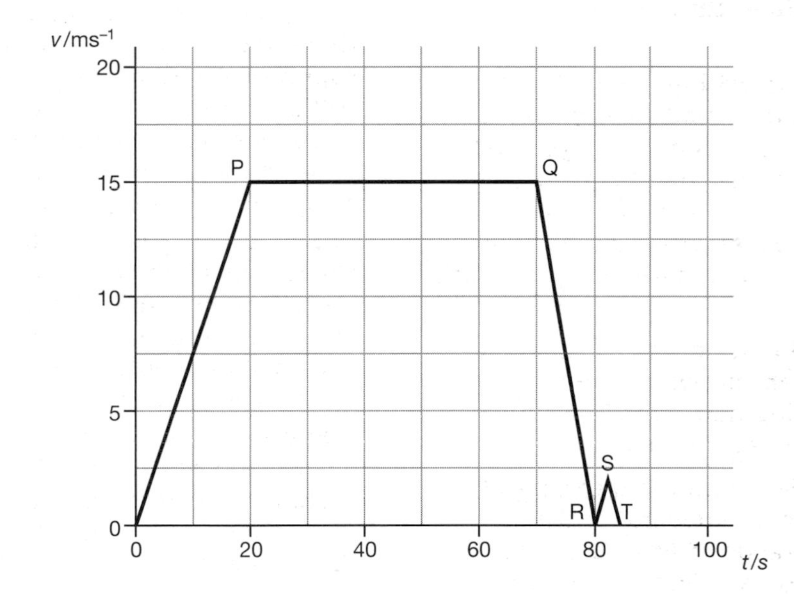

Calculate

(i) the acceleration of the car during the first 20 s of the journey,

$\frac{15}{20}$

acceleration = $0.75\ ms^{-2}$ ✓ ✓ — 2/2

(ii) the resultant force that acted on the car during this 20 s interval,

F = ma ✓

F = 800kg x 0.75 ms^{-2}

= 600 N

force = 600 ✓ N

[4 marks]

(iii) the distance travelled by car during the first 80 s of the journey.

$s = ut + \frac{1}{2}at^2$ ✗

$= 0 + \frac{1}{2} \times 0.75 \times 80s^2$

distance = 2400 m 0/4

[8 marks]

(c) On Fig. 2.1, RT indicates an interval of time during which the car was involved in a minor traffic accident. Suggest, with a reason, a likely nature of the accident.

When the car stops, a car from behind didn't stop quickly enough and hit the first car, the first car jogged forward a little during the hit, then stopped again. ✓ ✓ 2/2

[2 marks]

[Total 13 marks] 7/13

How to score full marks

(a) (i) This answer does not score any marks **because the symbols are not defined and the relation is inaccurate.** The answer implies that the student thinks that acceleration is velocity divided by time. In fact, acceleration is the change in velocity divided by time (or the rate of change of velocity). **You would be given full marks if you wrote the equation as a = (v–u)/t and defined each of the symbols.**

(b) (i) Although this answer gains full marks as it stands, it would be better if you stated that **the value of the gradient is equal to the acceleration**. A slip up in your calculation would otherwise mean that you scored zero here.

(iii) This fails to score any marks because it is incorrect to try to use an equation of motion when the acceleration is not constant. **To answer this, you must recognise that distance travelled is the area under the graph and calculate this. (975 m)**

(c) This is well answered because the answer **suggests what has happened** ('the car briefly moved forward and stopped again') and **gives a good reason for why it happened** ('the car was struck from behind by another car').

Don't make these mistakes ...

If a question asks you to **define something**, don't just write an equation. A definition should be in words. If you base your definition on an equation, **you must define each of the terms in the equation. This is a simple way to focus your definition.**

In calculations, **be sure to say what you are doing.** Simply stating that distance travelled is the area under a speed time graph is likely to gain you the first mark.

Make sure that you answer the question. **If you are told to give a reason then you must do so.** Simply stating the outcome won't get you full marks.

Key points to remember

Learn the two equations defining **average speed** and **acceleration** and you will be able to derive the other equations by substitution.

average speed = distance/time $\frac{s}{t} = \frac{v+u}{2}$

acceleration = change in velocity/time $a = \frac{v-u}{t}$

s = distance or displacement/m
t = time/s
u = initial speed or velocity/(ms^{-1})
v = final speed or velocity/(ms^{-1})
a = acceleration/(ms^{-2})

- Many examination questions on **motion** are expressed graphically and so it is essential that you can relate s, v and a at specific times.
- **Gradients and areas under the line** are the only quantities that are likely to be measured for graphs; by considering the units of the gradient or the area, you can decide on the quantity that is being measured.
- For a **distance/time graph** the gradient(s) is in ms^{-1} and indicates speed (the area has no physical significance).
- For a **speed/time graph** the gradient(s) is in ms^{-2} and indicates acceleration; area is in m and indicates distance travelled.

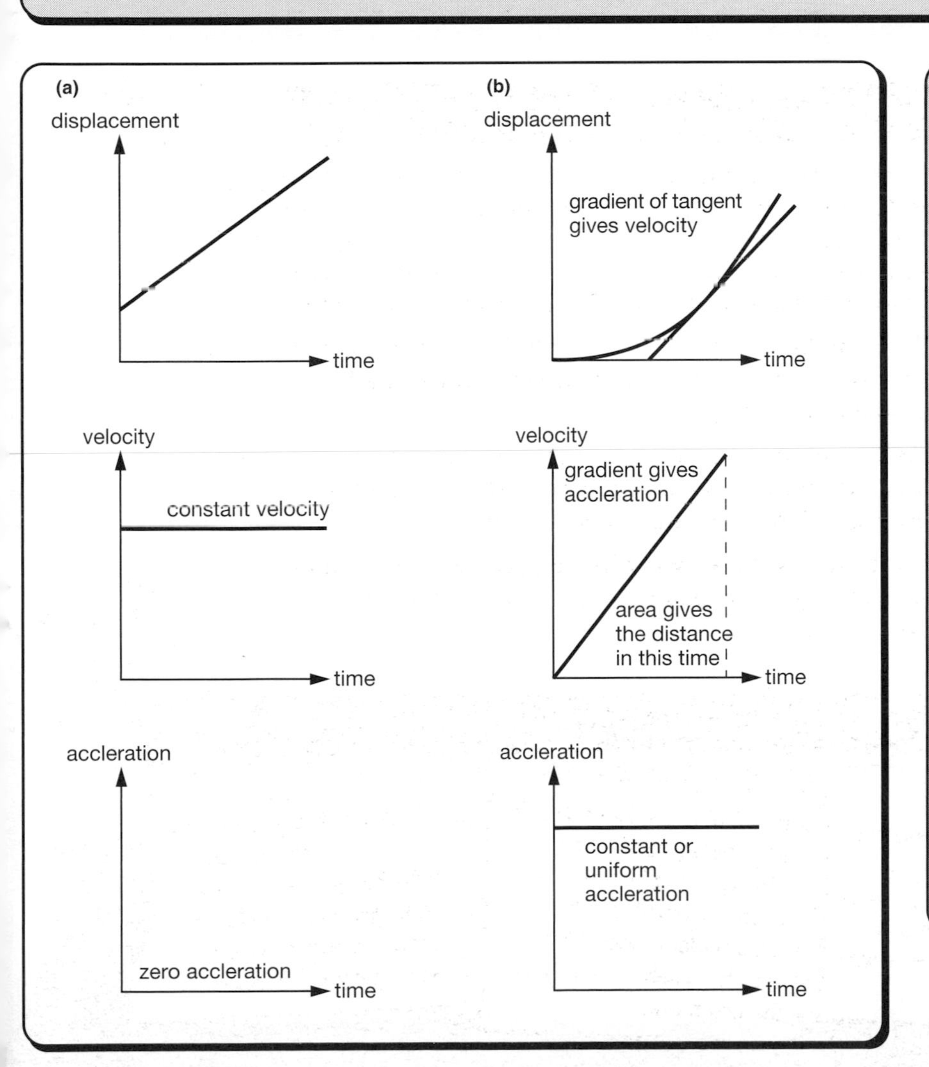

Diagrams **(a)** show how **displacement, velocity and acceleration are related** (when there is zero acceleration and no resultant force acting): displacement changes uniformly with time (but doesn't start from zero) so the velocity is constant; the velocity does not change (zero gradient) and so there is no acceleration.

Diagrams **(b)** show a **constant (non-zero) acceleration** which shows that the velocity increases linearly with time (constant gradient). The gradient of the displacement-time graph varies and so it is essential to measure the gradient of the tangent at a particular time – these are the graphs that would be produced by a free-falling object.

Questions to try

Examiner's hints for question 1

- When you answer a question based on a graph, you must think hard about the units of the area and the gradient. These will give you a good idea about what you are calculating e.g. $ms^{-1}/s = m/s^2$ (or ms^{-2}) which tells you the gradient gives acceleration.
- Write down the equations for acceleration and average speed and see how each relates to the graph.

Fig. 1.2 shows how the velocity of an object varies with time.

Fig. 1.2

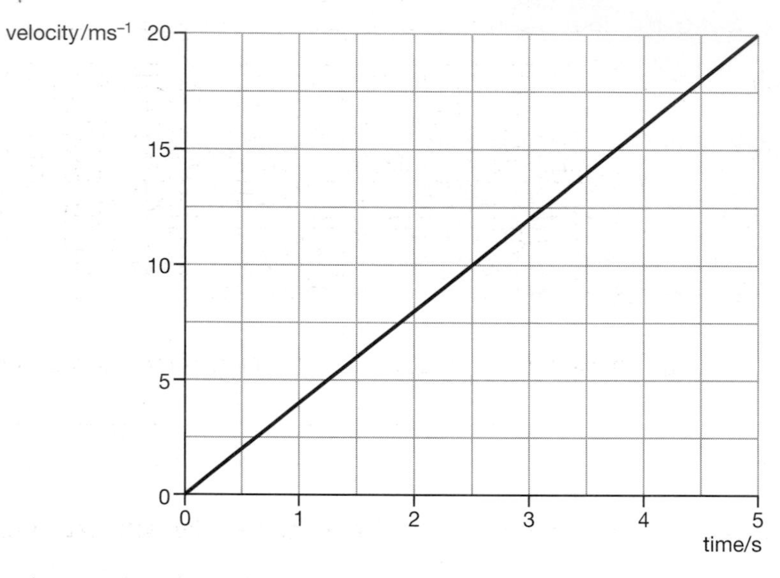

(a) Calculate the acceleration of the object.

[2 marks]

(b) Calculate the distance that it travels in 4 s.

[2 marks]

[Total 4 marks]

Examiner's hints for question 2

(a) You must think about what is happening to the direction of the velocity as the ball hits the ground and what the scale division is for time.

(b) Think carefully about the units of slope.

(c) Don't forget that height is a distance (given by the area under the line). With a 'show' question you are expected to provide clear, detailed working.

(d) Try to concentrate on the key aspects of the motion. You will be expected to include values of displacement. Decide when the ball hits the ground and what its displacement must be from the point of release. Remember that the ball is accelerating and so it will move through greater distances in later portions of time (until it hits the ground) – and so the line will not be linear.

(e) Don't forget that displacement is being measured from the point of release and it is a vector (see chapter 2).

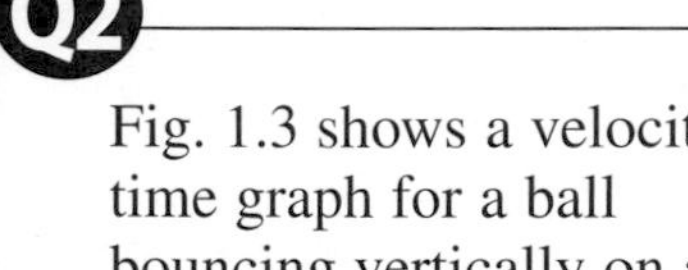

Q2

Fig. 1.3 shows a velocity-time graph for a ball bouncing vertically on a hard surface. The ball was dropped at t = 0 s

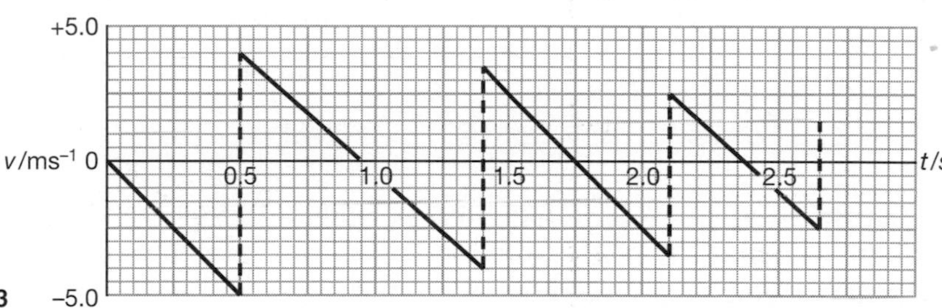

Fig. 1.3

(a) At what time does the graph show the ball in contact with the ground for the third time?

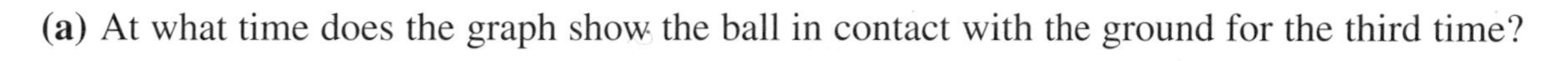

....

[1 mark]

(b) The downwards-sloping lines on the graph are straight and parallel with each other. Why?

....

....

[2 marks]

(c) Show that the height from which the ball is dropped is about 1.2 m.

....

....

[2 marks]

(d) Sketch a displacement-time curve on the axes below for the first second of the motion.

[3 marks]

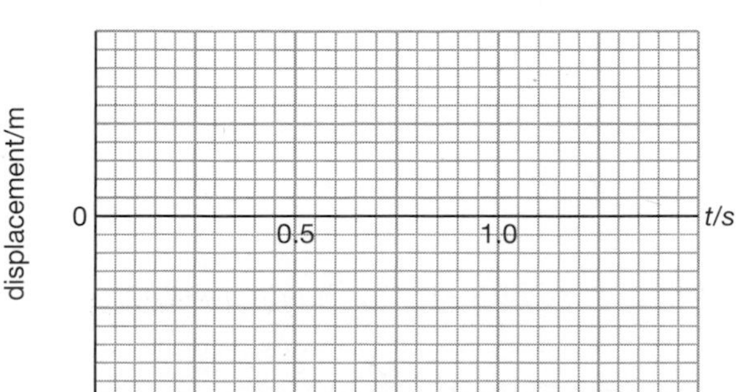

(e) What is the displacement of the ball when it finally comes to rest?

....

[1 mark]

[Total 9 marks]

The answers to these questions are on page 85.

Exam question and student's answer

1 (a) State the important difference between a vector and a scalar.

Vector has direction ✓ (1/1)

[1 mark]

(b) Put the following quantities into a list of vectors and a list of scalars.

[3 marks]

MASS FORCE SPEED VELOCITY WORK DISPLACEMENT

vectors	scalars	
velocity	*mass*	✓
displacement	*speed*	✓
force	*work*	✓

(3/3)

(c) In order to display greetings cards, a student fixes a length of string between two nails and then suspends the cards from the string. Fig.2.1 shows the string with one card of weight 0.60 N suspended by a light clip at the centre of the string.

Fig. 2.1

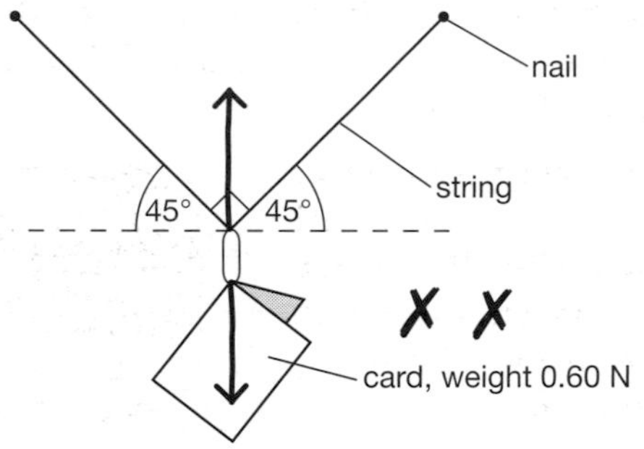

nail
string
45°
45°
card, weight 0.60 N

✗ ✗

(i) On Fig.2.1, mark the forces on the clip due to the tension in the string.

(ii) The resultant of the forces due to the tension in the string is 0.60 N. In the space below, draw a vector triangle for the forces in the string and their resultant. Use a scale of 1.0 cm to represent 0.10 N.

Fig. 2.2

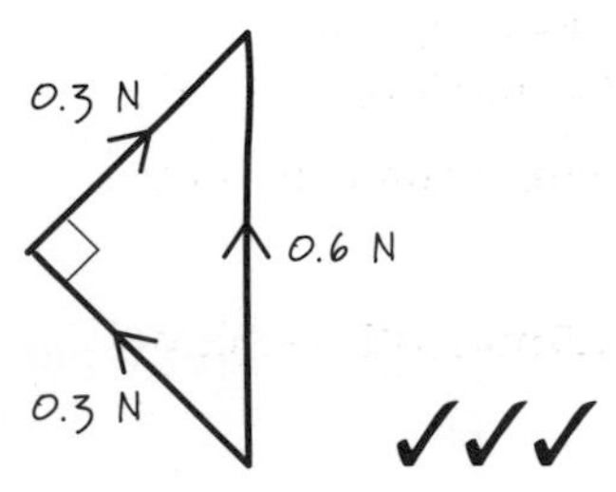

✓✓✓ (3/3)

(iii) Use your completed vector diagram to determine the magnitude of the tension in the string. (0/1)

tension = *0.3 N* ✗ N [6 marks]

(d) The student decides that she would like as little sag as possible in the string when it is loaded with cards. To achieve this, she tightens the string. State, with a reason, whether the string, loaded with cards, could ever be horizontal.

No. There is no directly opposite force to the card thus there is only a downward force on the string which cannot be compensated for by the tension alone. ✓ 1/2

[2 marks] 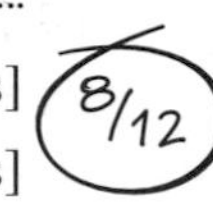

[Total 12 marks]

How to score full marks

Part (a) (i)

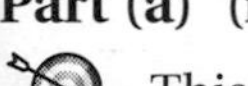

This gains the mark as it stands, although **you could make the difference clearer** if you added "....scalars have no direction".

Part (b)(i) gains full marks as it stands.

Part (c)(i)

This does not score any marks. The two forces marked on the clip are the weight of the card and the **resultant** pull of the string that balances the weight of the card. Despite these being correctly drawn, the student is not answering the question asked. **The tension acts in the direction of the string and should be shown** as:

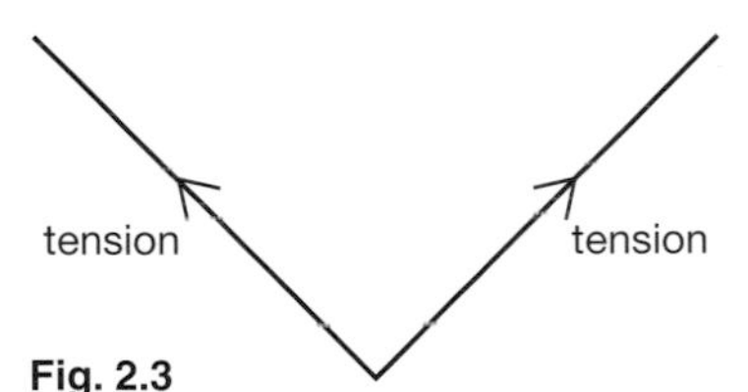

Fig. 2.3

To make your answer completely clear, with no doubt about what you mean, you would do well to label the tension.

Part (c)(ii)

This gains all three marks because the diagram is correct. Error in the labelling of the values of the two tensions is penalised in c(iii).

Part c(iii)

The student **has not measured the length of the diagonal lines with a ruler** and so has lost this mark.

Pythagoras' theorem shows that this cannot be correct:

$(0.3^2 + 0.3^2$ cannot $= 0.6^2)$.

The correct answer is 0.43 N (21.5 mm with a scale of 5 mm:0.1 N).

Part (d)

This is nearly correct **but it is not clear enough to allow both marks to be awarded**. There is still doubt about what the answer means. You need to write "there must be a vertical component of the tension to equal the weight; when the string is horizontal it cannot have a vertical component." You would not get a mark for simply stating that the string could never be horizontal.

Don't make these mistakes ...

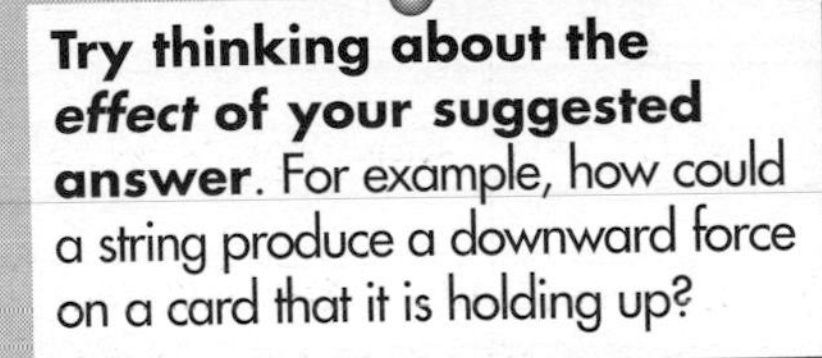

Try thinking about the *effect* of your suggested answer. For example, how could a string produce a downward force on a card that it is holding up?

Think your way through calculations – 0.3 N added to 0.3 N could only give a resultant of 0.6 N if they were in the same direction.

In descriptive questions, apply your knowledge of physics clearly. Practise asking yourself, 'Does my answer actually say what I mean it to?' 'How could I make it clearer?'. It is only by writing answers to descriptive questions and then getting others to interpret them that we realise that the meaning of what we write is often not as clear as it needs to be.

Key points to remember

- **Scalars** are quantities with size alone: mass, length, time, speed, density, pressure and energy.
- **Vectors** are quantities with size and direction: velocity, displacement, acceleration and force.

- **Scalars** are treated as ordinary numbers when you add, subtract, multiply or divide.
- **Vectors** require vector algebra when you add, subtract, multiply or divide them – you only need to be able to add and subtract.

- When two vectors are added the combination of them is called the **resultant**.
- Vectors may be added by using a **scale diagram** (don't forget to include the scale) and the resultant measured directly (and then scaled).
- Vectors may also be combined by calculating the resultant from a **sketch diagram**. Use Pythagoras' theorem or trigonometry (cos, sin and tan).

When vectors are combined they must be added so that one follows on from the other in the direction of the arrows:

each line shows the direction of the vector and has a length proportional to the magnitude of the vector

vector A

vector B

vector C

vector C = vector A + vector B

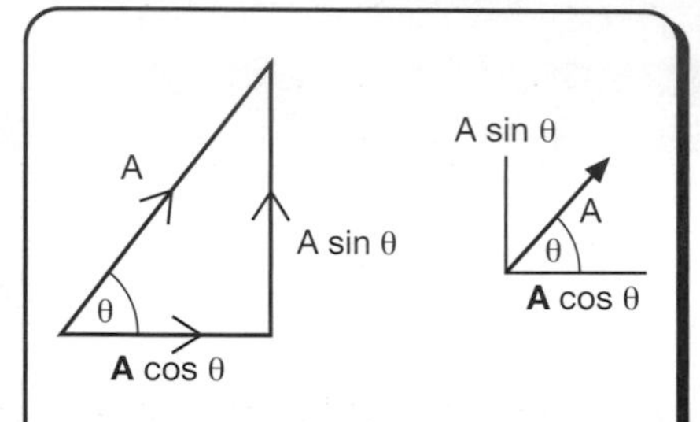

A has been resolved into a horizontal component (= **A** cos θ) and a vertical component (= **A** sin θ).

- Often it is convenient to split vectors up into two **perpendicular components** – this is known as resolving them into components.
- **Perpendicular components** can be treated separately e.g., in projectile motion the **horizontal** component of the velocity stays constant while the **vertical** component of the velocity changes under the influence of the gravitational field.

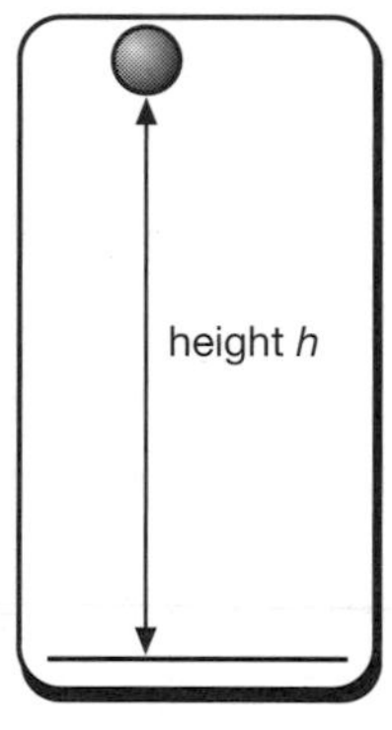

- Timing a falling object (using a timer, light gates or an electromagnet and mechanical gate) for different heights allows the acceleration of free fall, and hence the **gravitational field strength**, to be calculated graphically.
- A graph of h against t^2 will be of gradient $g/2$.

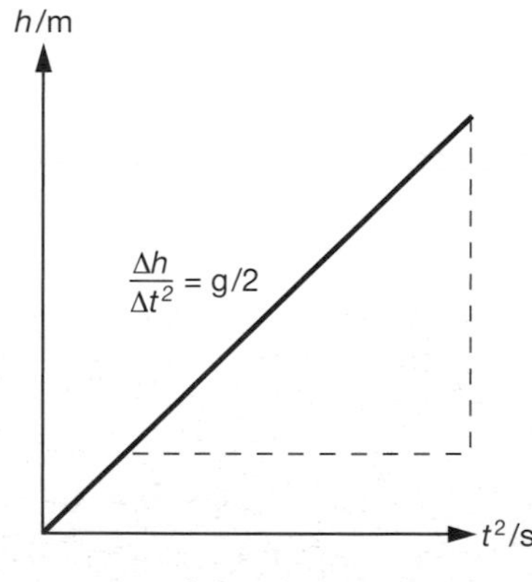

- Be careful with units here.
- Make sure your gradient triangle is large.
- Double the gradient gives g.
- For accurate readings heights should be as large as possible.
- Don't forget to repeat and average all readings.

- **Projectile motion** is dealt with by considering the horizontal and vertical motion separately.

  The horizontal velocity does not change throughout the time considered (when we ignore air resistance) – the distance travelled will be given by:

  $s = u_h \times t$ (u_h is the horizontal velocity and t is the time).

  The vertical velocity is changed constantly by the effect of the gravitational field – the height fallen is given by :

  $h = u_v t + \frac{1}{2} g t^2$ (u_v is the vertical velocity, g is the gravitational field strength and t is the time).

  Whether objects accelerate or decelerate depends on the relative directions of the velocity and the gravitational field.
- **Air resistance** (drag) increases the time taken for the projectile to rise or fall.

Questions to try

Examiner's hints for question 1

(a) Since the words scalar quantity and vector quantity are italicised, you are expected to mention each quantity.

(b) Be careful with units here. You must use consistent units, although it is not necessary to convert all times into seconds.

Q1

(a) Distinguish between a *scalar quantity* and a *vector quantity*.

....

....

....

....

[2 marks]

(b) A car travels one complete lap around a circular track at an average speed of 100 km h^{-1}.

(i) If the lap takes 3.0 minutes, show that the length of the track is 5.0 km.

....

....

....

(ii) What is the magnitude of displacement of the car after 1.5 minutes?

....

[4 marks]

[Total 6 marks]

Examiner's hints for question 2

(a) and (b) Look at the marks available for each of these parts. The answer to **(b)** is likely to be much more straightforward than the answer to **(a)**.

(a) The axes for velocity are both positive and negative and velocity is a vector. This should make you realise that you need to think about both positive and negative velocities. When the ball is first released it starts from rest. At the time that it bounces, its direction changes. Since it is falling under gravity, the acceleration of the ball is constant. Acceleration is given by the gradient of a velocity/time graph.

(b) The value of acceleration due to gravity is 9.8 ms^{-2}

(c) You must label the diagram clearly. This makes your meaning entirely obvious to the examiner. Gravity always acts vertically downwards. The direction of the momentum will be that of the velocity. The direction of the momentum is that of the velocity (since momentum = mass × velocity).

(d) First you need to find the horizontal component of the velocity. Next, to get the momentum, you must multuply this by the mass. Finally, recognise that this will not change since there are no external forces acting on the ball (we ignore air resistance). You would usually be expected to give a direction for the momentum – it's a vector. In this case you are asked for the magnitude of the momentum and so the direction is not required.

A ball is dropped and rebounds vertically to less than the original height.
For this first bounce only, sketch graphs of

(a) the velocity of the ball plotted against time,

(b) the acceleration of the ball plotted against time.

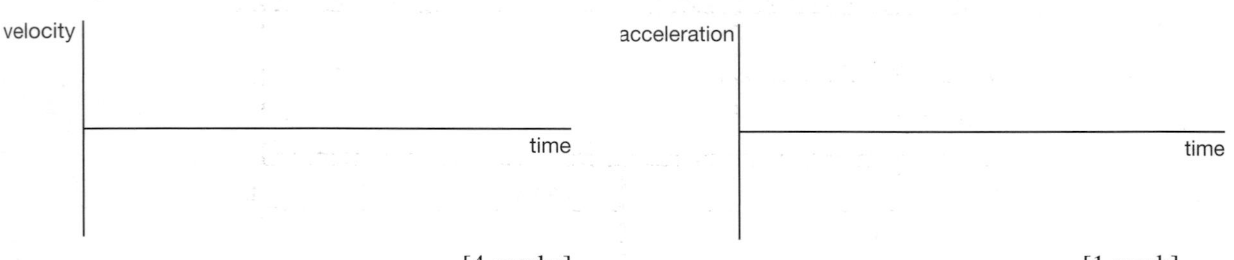

[4 marks] [1 mark]

(c)

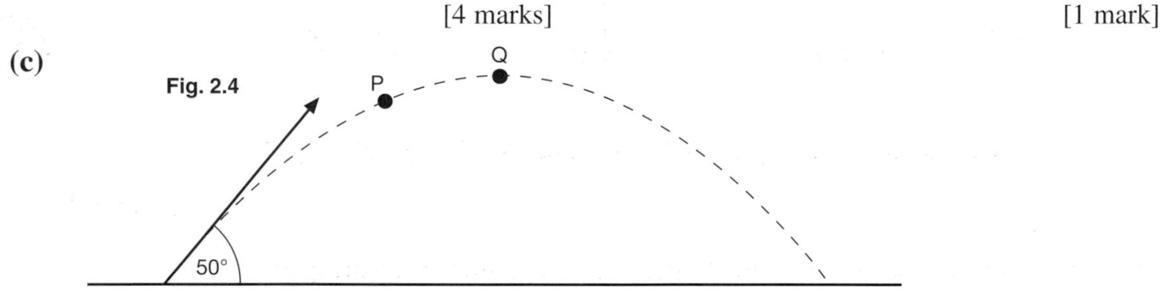

Fig. 2.4

The ball is then thrown at an angle to the horizontal and follows the trajectory shown in the diagram.

Mark on the diagram the directions of

(i) the acceleration vector at P,

(ii) the acceleration vector at Q,

(iii) the momentum vector at P,

(iv) the momentum vector at Q.

[4 marks]

(d) The mass of the ball is 0.15 kg and the initial direction makes an angle of 50° to the horizontal. Calculate the magnitude of the momentum of the ball at Q when it is projected with an initial speed of 15 ms^{-1}. Neglect the effects of air resistance.

....

....

....

The answers to these questions are on pages 85 and 86.

[4 marks]

[Total 13 marks]

Exam question and student's answer

(a) Define the moment of a force about a point.

Force x perpendicular distance ✓

[2 marks]

1/2

(b) The system shown in Fig 3.1 is in equilibrium. The uniform rod, CD, has a weight of 15 N and is suspended by two lengths of string A and B.

Fig. 3.1

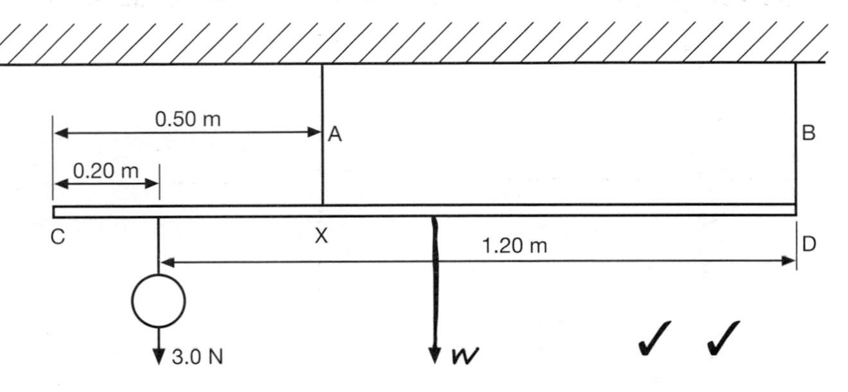

2/2

(i) Accurately draw an arrow onto Fig 3.1 to show the weight of the rod.

[2 marks]

(ii) By taking moments about point X, determine the tension in string B.

$$(3\text{N} \times 0.3\text{m}) + B \times 0.9\text{m} = 15\text{N} \times 0.2\text{m}$$
$$0.9\text{Nm} + (B \times 0.9\text{m}) = 3\text{Nm}$$
$$B \times 0.9\text{m} = 2.1\text{Nm}$$
$$B = 2.33\text{N}$$ ✓✓

2/2

[2 marks]

(iii) Determine the tension in string A.

$$2.33 + T_A = 3 + 15$$
$$T_A = 15.67\text{N}$$ ✗ significant figure penalty

[1 mark]

[Total 7 marks] 5/7

How to score full marks

Part (a)

This answer fails to score full marks because **the student has not stated that the *perpendicular* distance is measured to the point.**

Part (b) (ii)

The student has recognised that the tension in string **B** produces a moment in the same direction as that produced by the 3.0 N weight (anticlockwise). **This answer is well set out with 'B' representing the tension in string B.**

Part (b) (iii)

The answer is correct but **the student has been penalised for using an unrealistic number of significant figures.** The data is all given to two or three significant figures and so your answer must be given to the same precision. The answer should be either 15.7 N or 16 N.

Don't make these mistakes ...

In definitions **do not write answers that contain any form of short hand**. You might understand what the shorthand represents, but the examiners will not give you the benefit of the doubt if your answer is at all unclear.

Do set out your calculations clearly. This makes the examiner's task easier and will count in your favour!

Be careful with significant figures. Examiners will not expect you to use too many or too few. It is a good idea to base your number of significant figures on the least number of significant figures given in the data. **Even easier – you will not usually be penalised for using two or three significant figures.**

Key points to remember

Forces are physical quantities that fulfil the conditions set by Newton's laws. Resultant (unbalanced) forces:

- change the motion of an object by either:

  changing the speed (possibly from rest).
  or changing the direction of the motion.
- produce a rate of change of momentum (mass x velocity) of the object, to which they are applied, which will be proportional to the size of the force.
- occur in pairs such that the two forces are equal in size, opposite in direction and act on different bodies.

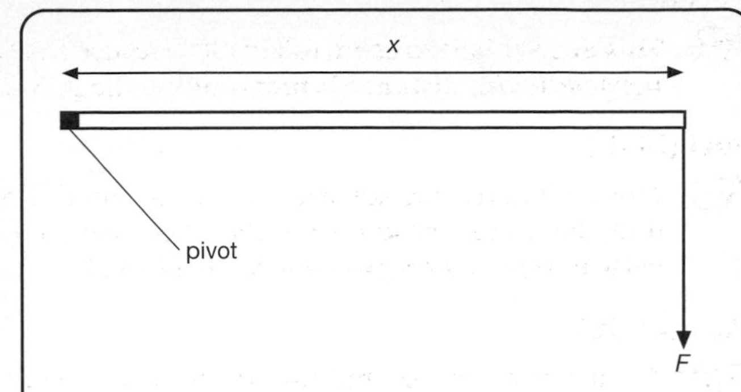

The moment of the force F about the pivot is given by the product Fx.

This moment is clockwise and is unbalanced and so would cause the object to rotate.

Newton's second law is often written in the form:

$$F = ma$$

This is fine if the mass does not change (which is usually the case). It shows us that the unit N is equivalent to kgms^{-2}.

Don't forget that moments are measured in units of Nm (or Ncm etc.) NOT in N/m

The **principle of moments** states that for an object in equilibrium, the sum of the clockwise moments equals the sum of the anticlockwise moments.

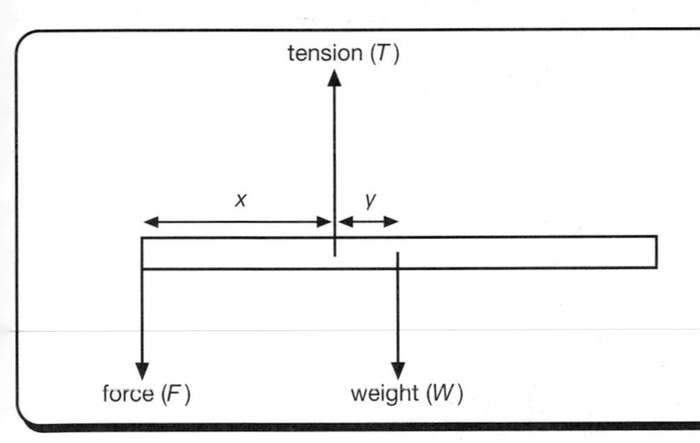

The diagram shows a ruler hanging from a string. Its weight, w, is being counter-balanced by the force, F. By taking moments about the string we cancel the effect of the tension so:

$$Fx = Wy$$

We can also equate the vertical forces and so:

$$T = W + F$$

When we measure F, x and y we can calculate both the tension and the weight of the ruler.

When an object is in **equilibrium**, there is no resultant force or resultant moment acting on it. This means that there is a closed polygon of forces that all act through the same point to give no net moment.

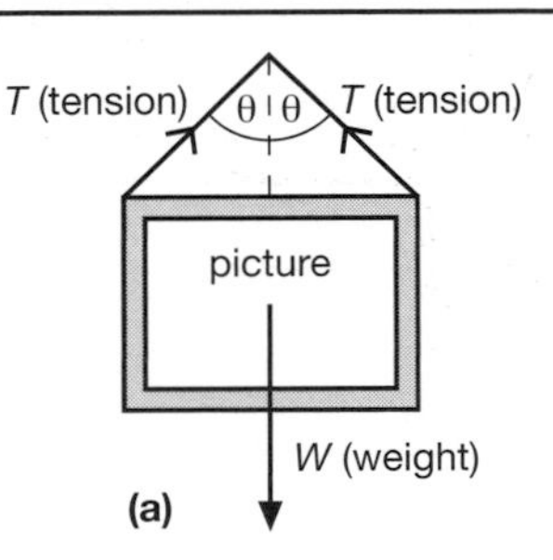

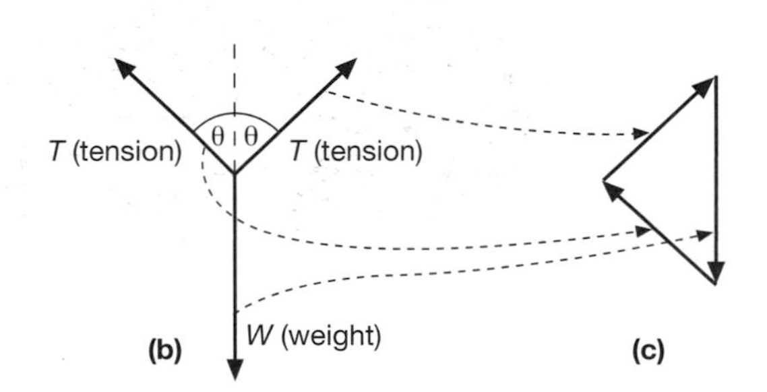

(a) shows the forces acting on a picture in equilibrium.

(b) shows how the forces can be arranged to act through a single point. In this case, the forces cancel each other out.

(c) shows the forces arranged in a closed triangle, indicating they are in equilibrium. Usually, a scaled diagram of this would be a good way to answer to the questions on equilibrium.

Questions to try

Examiner's hints for question 1

- Think about the difference between mass and weight and how to convert from one to the other.
- When resolving forces vertically, don't forget to use cosine when the angle is measured from the vertical and sine when the angle is measured from the horizontal. Be careful in this question because there is a mixture of horizontals and verticals.
- In the final discussion, don't just make a statement. The word 'discuss' tells you that you should justify any statement(s) you make.

The forces acting on the hands and feet of a rock climber change continually as the climber moves across the rock face.

A climber of mass 70 kg, who is carrying a rucksack of mass 20 kg, is in the process of moving his right foot to new toehold. In the diagram he is temporarily at rest whilst looking for a suitable position for his foot

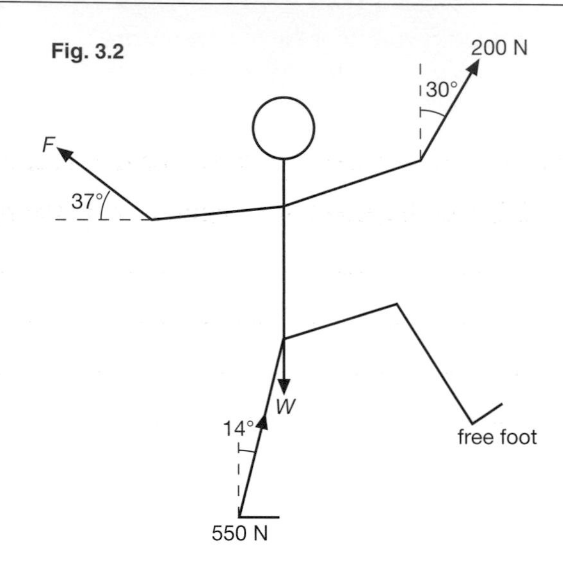

(a) What is the total weight W of the climber and his load?

....

....

$W=$ ....

[2 marks]

(b) Show that the vertical components of the 200 N and 550 N forces add up to approximately 700 N.

....

....

....

....

[2 marks]

(c) The climber is in equilibrium. What do you understand by the term *equilibrium*?

....

....

[1 mark]

(d) Calculate the magnitude of force F.

....

....

....

....

$F =$ ....

[2 marks]

(e) The climber sees a suitable position and accelerates his free foot vertically upwards.

Discuss the effect this might have on the force exerted on the other foot.

....

....

....

....

[2 marks]

[Total 9 marks]

Examiner's hint for question 2

- The key aspect of this calculation is that the moment is the product of the force and the perpendicular distance from the direction of the force to the point. You should mark this distance on the diagram to help you to calculate it.

Q2

(a) (i) Define the moment of a force. Illustrate your answer with a diagram.

....

....

....

(ii) Define torque of a couple.

....

....

....

[4 marks]

(b) An electricity cable is attached to a pole at a height of 6.0 m above the ground as shown in Fig 3.3.

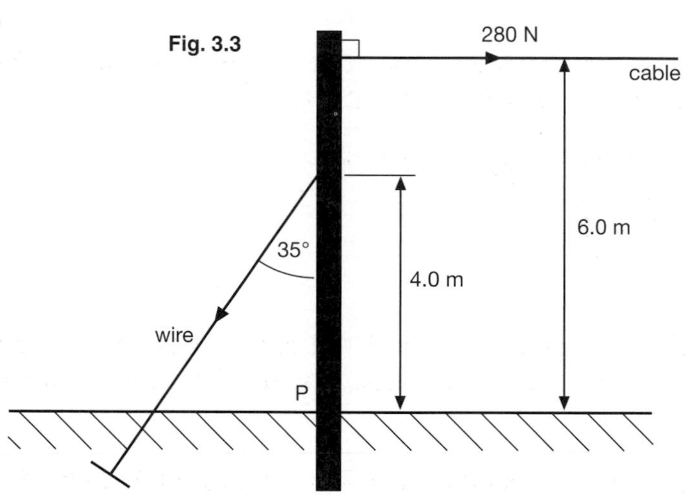

The cable exerts a force of 280 N on the pole at an angle of 90° to the pole. So that there is zero turning moment on the pole itself, a wire under tension is attached to the pole at a height of 4.0 m and it makes an angle of 35° to the pole.

Calculate

(i) the moment which the cable exerts about P, a point in the pole level with the ground.

moment = ... N m

(ii) the tension necessary in the wire.

tension = ...N

[5 marks]

The answers to these questions are on pages 86 and 87. [Total 9 marks]

Exam question and student's answer

Palaeontologists are able to deduce much about the behaviour of dinosaurs from the study of fossilised footprints.

The tracks below show the path of the *Tyrannosaurus Rex* as it attacks a stationary *Triceratops*.

Fig. 4.1

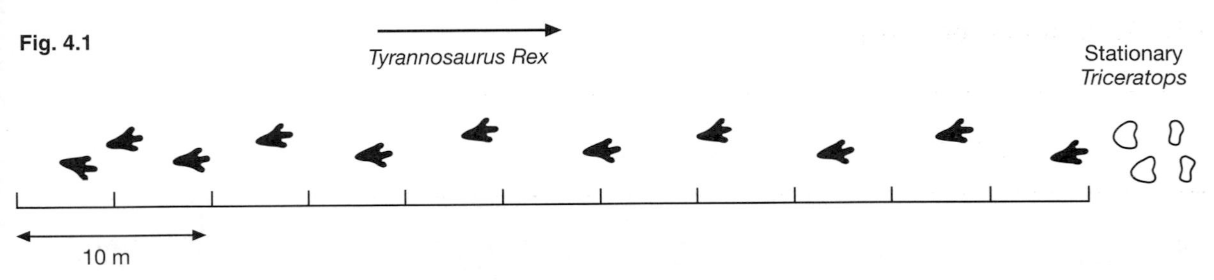

(a) The time between footprints is 0.62 s. Show that the maximum speed of the *Tyrannosaurus Rex* is about 10 ms^{-1}.

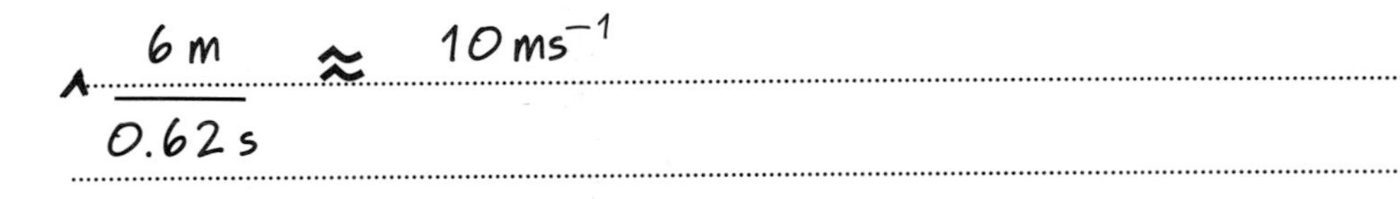

[2 marks]

(b) *Tyrannosaurus Rex* is believed to have attacked its prey by charging and locking its jaws on the prey. *Tyrannosaurus Rex* would be at its maximum speed when it hit the stationary prey.

(i) This *Tyrannosaurus Rex* has a mass of 7000 kg. Calculate its momentum just before it hits the *Triceratops*.

$p = mv = 7000 \times 10 = 70000$ ✓

Momentum = 70000 N ✗

[2 marks]

(ii) *Triceratops* has a mass of 5000 kg. Calculate their combined speed immediately after the collision.

$70000 = (7000 + 5000)\,V$ ✓ ✓

$V = 5.83\ ms^{-1}$

Combined speed = $5.83\ ms^{-1}$ ✓

[3 marks]

(c) The skull of *Tyrannosaurus Rex* is heavily reinforced to withstand the force produced in such a collision.

Calculate the force exerted on the *Tyrannosaurus Rex* if the time taken to reach their combined speed after the collision is 0.30 s.

$F = \frac{\text{momentum}}{\text{time}}$ ✗

$= \frac{7 \times 10^4}{0.30}$

$= 2.3 \times 10^5$ N ✗

Force = 2.3×10^5 N ✗ 0/3

[3 marks]

[Total 10 marks] 4/10

How to score full marks

Part (a)

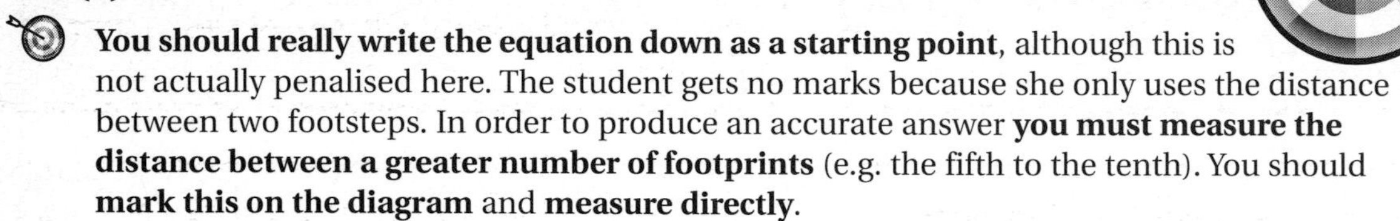

You should really write the equation down as a starting point, although this is not actually penalised here. The student gets no marks because she only uses the distance between two footsteps. In order to produce an accurate answer **you must measure the distance between a greater number of footprints** (e.g. the fifth to the tenth). You should **mark this on the diagram** and **measure directly**.

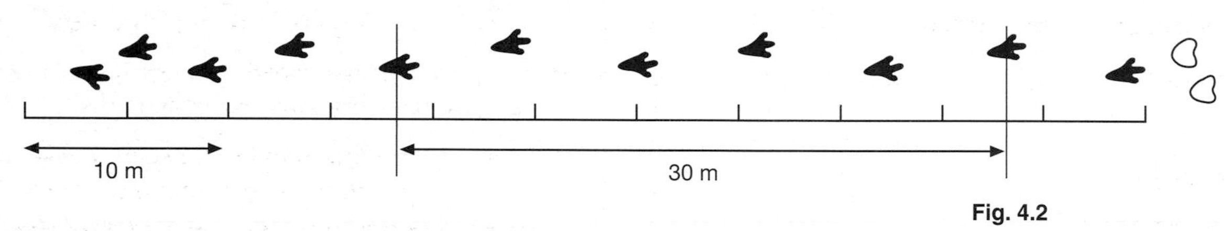

Fig. 4.2

v = 30 m/ (5 × 0.63 s) = 9.7 ms^{-1}

The student has rounded the number to an integer and has lost a mark because of **not being plain and precise enough.**

Part (b) (i)

This part of the question is nearly completely correct but the unit of momentum is wrongly written as *N* – it should be **Ns** or **kg ms^{-1}**.

Part (b) (ii)

This answer is completely correct but again **the answer would have been better if the student had written the equation.**

Part (c)

This is completely wrong! **Using an incorrect equation means that the reasoning in the rest of the answer is also wrong** and no marks are earned. The equation should be:

$F = \frac{\Delta(mv)}{\Delta t}$ or force = **change** in momentum per second

so F = 7000 kg × (10 ms^{-1} – 5.8 ms^{-1})/0.30 s [*using the student's values*]

F = 97 000 N (to two significant figures)

Don't make these mistakes ...

In graphical and other questions where you need to take a reading from a scaled diagram, take the largest reading possible. This is important because making a small error when reading a large quantity is much less significant than the same small error in the reading of a small quantity. **Take particular care when choosing a gradient triangle on a graph – make it large!**

Don't forget to annotate your diagram to help explain what you are doing. **It is your examination paper and you are allowed to do whatever you think necessary to help your explanations.**

Key points to remember

- **Work** is done when a force moves an object.
- Work is defined as the product of the force and the distance moved in the direction of the force.
- $W = Fs$
- Work is measured in units of Nm or J.
- Work is a scalar.

- The two main types of **mechanical energy** are **kinetic** and **potential**.
- $E_K = {}^1/_2\, mv^2$
- $E_G = mgh$
- $E_E = {}^1/_2\, kx^2$

E_K = kinetic energy
E_G = gravitational potential energy
E_E = elastic potential energy
k is the stiffness of a spring (force per unit extension)
x is the extension of the spring

- **Power** is the work done per second or the energy transferred per second.
- $P = \dfrac{\Delta W}{\Delta t}$
- Power has units of Js^{-1} or W.
- Power is a scalar quantity.

- **Energy** is the ability to do work
- There are numerous names given to energy. These are based on the way energy is transferred but all energy fits in with the above definition.
- If we consider everything then energy is always conserved.

- **Conservation of energy** applies whenever all forms of energy are taken into account.
- Energy cannot be created or destroyed, merely changes from one form to another.
- When we consider nuclear changes we see that this is really conservation of mass/energy.

- Many questions can be answered by using the laws of conservation of energy. It is one of the key tools that you should consider when deciding which method to use to answer questions.

- Efficiency = $\dfrac{\text{useful energy output}}{\text{total energy input}} \times 100\%$
- Energy is *never* lost; it is simply converted into a form that may not be as useful to us.

Key points to remember

Conservation of momentum. This applies to all interactions providing that no external force is acting.

The sum of the momentum after an interaction = the sum of the momentum before the interaction.

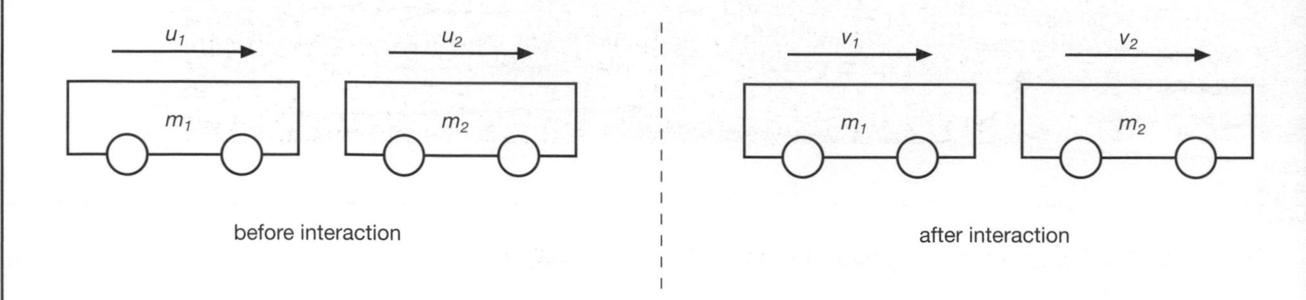

$m_1v_1 + m_2v_2 = m_1u_1 + m_2u_2$

m_1 and m_2 are masses in kg, u_1 and u_2 are initial velocities in ms^{-1} and v_1 and v_2 are final velocities in ms^{-1}.

Collisions

- In **elastic collisions**, conservation of momentum and kinetic energy applies. Some molecular collision are elastic.
- In **inelastic collisions**, conservation of momentum applies but some kinetic energy is transferred to other forms. Balls bouncing on the ground experience inelastic collisions.
- In **totally inelastic collisions**, conservation of momentum applies but the colliding objects stick together. An example of this is warm plasticine striking the floor.
- In an **explosive collision**, conservation of momentum applies but there is a gain of kinetic energy (from other forms). Examples include an arrow being released from a stretched bow string or the emission of an α-particle.

Impulse

Impulse is found by multiplying the **force** by the **time** for which it acts or finding the **area** under a force against time graph.

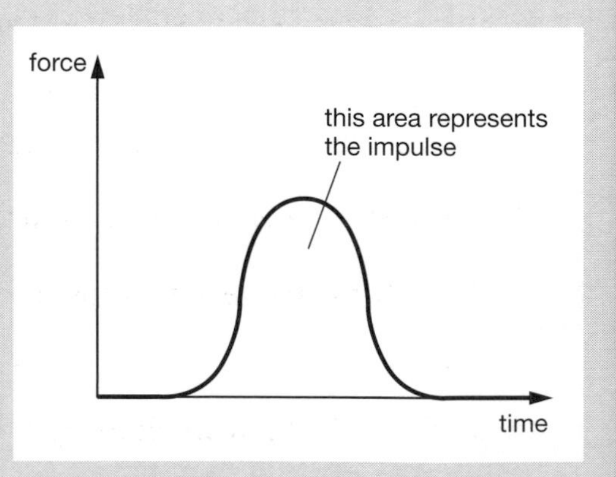

- Impulse = change in momentum
- $Ft = mv - mu$
- Impulse is measured in Ns and so this can be used as an alternative to the kg ms^{-1}, the unit of momentum.
- Impulse and momentum are both **vector quantities**.
- This has many implications in sport and for transport safety. For example, when a car brakes the passenger continues to move with the velocity of the car. By slightly stretching, the seatbelt prolongs the time for which the force acts and so the momentum is brought to zero with a smaller force thereby reducing the risk of injury.

Question to try

Examiner's hints for question 1

- The word varying is crucial in **part (a)** and means that you cannot simply define work as force x distance.
- In **part (b)(i)** you must think about whether force is a vector or a scalar quantity.
- **Part (b)(ii)** is really asking what happens to the amount of kinetic energy when a negative force is acting.
- The key to **part (d)** is the phrase 'constant speed.'

Q1

(a) State in words how to calculate the work done by a varying force.

....

.... [2 marks]

(b) (i) Under what circumstances is the work done by a force negative?

....

(ii) What happens to the kinetic energy of the body on which the force acts in such circumstances?

....

.... [2 marks]

A runaway sledge slides down a slope at a constant speed. One force is shown on the free-body diagram of the sledge. It is the normal contact push of the snow on the sledge.

Fig. 4.3

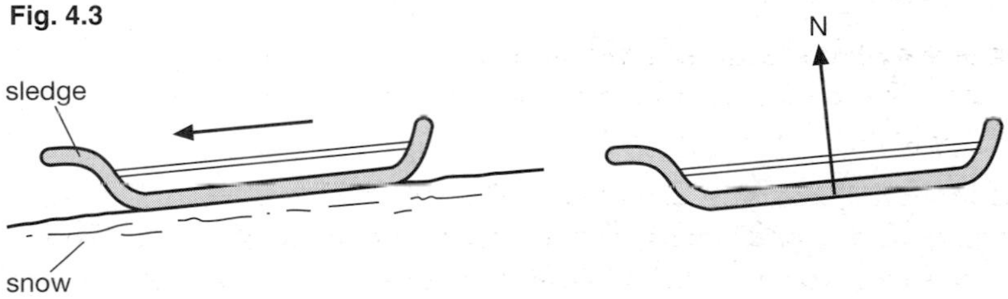

(c) Add to the free-body diagram to show the other two forces acting on the sledge. Name each force and state what is producing it.

....

.... [3 marks]

(d) The sledge slides 15 m down the slope at a constant speed. The force $N = 40$ N.

(i) What is the resultant force acting on the sledge?

....

(ii) What is the work done by the force N?

.... [2 marks]

[Total 9 marks]

The answer to this question is on page 87.

Exam question and student's answer

1 A metal wire of length l and area of cross section A is fixed at one end and hangs vertically with a load W attached to its free end. The wire is found to stretch by an amount Δx.

(a) Give, in terms of l, W and Δx, expressions, one in each case, for

(i) stress,

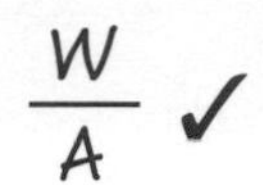

$\frac{W}{A}$ ✓ (1/1)

(ii) strain,

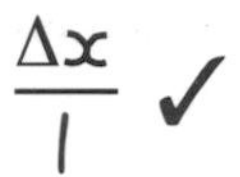

$\frac{\Delta x}{l}$ ✓

(1/1)

(iii) the Young modulus of the metal.

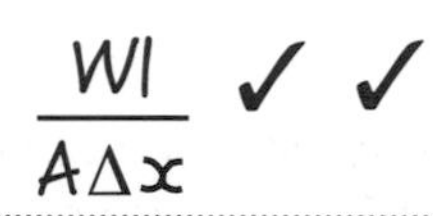

$\frac{Wl}{A\Delta x}$ ✓ ✓ (2/2)

[4 marks]

(b) The wire has length 2.5 m. A tensile stress of 6.4 x 10^7 Pa is applied. The Young modulus of the metal is 1.1×10^{11} Pa. For the wire, calculate

(i) the strain,

$$\frac{6.4 \times 10^7}{1.1 \times 10^{11}} = 5.82 \times 10^{-4}$$ ✓ ✓

strain = 5.82×10^{-4} (2/2)

(ii) the extension.

$$5.82 \times 10^{-4} \times 2.5 = 2.3 \times 10^{-3}$$ ✓ ✗

extension = 2.3×10^{-3} m

(1/2)

[4 marks]

(c) Suggest, with a reason,

(i) whether a 30 cm rule would be a suitable measuring instrument for the extension,

No, it would not be precise enough ✓

(1/2)

(ii) what would happen to the extension if the wire were to be replaced by another wire of the same dimensions made of a metal with a smaller Young modulus. Assume that the load remains the same and that the wire does not exceed its elastic limit.

There would be a greater extension ✓ (1/2)

[4 marks]

[Total 12 marks] (9/12)

How to score full marks

Part (a)

 The student's answer is correct and gains full marks. **The answer could be improved by making each of the parts into an equation**. In **part (iii) you should write the Young's modulus equation** ($E = \sigma/\varepsilon$, or Young's modulus = stress/strain). This would then act as a safety net in case your calculation was wrong.

Part (b) (i)

 Again this is correct but **without an equation it gives no room for error**.

Part (b) (ii)

 The first part of the calculation is correct but the student has gone on to **divide by 2.5** rather than **multiply by 2.5** and has also made an error with the power. The correct answer should be

1.46×10^{-3} m.

Part (c)(i)

 The statement that a ruler "is not precise enough" is correct and the student has used precise in the correct context. However, **only one mark out of the two possible is gained because there is too little information**. It is essential to support such a statement with further reasoning ... *"the ruler is accurate to ± 1 mm, but the extension is likely to be less than this."*

Part (c)(ii)

 Again the student has given the correct answer **but has not supported the statement**. It is **essential** to **state the reason for your answer in this type of question**... *"$E = \sigma/e$, so when E decreases with a constant stress, the strain must increase and so the extension must increase."*

Don't make these mistakes ...

Don't confuse the word **precise** with the word **accurate**.

A measurement is precise if you are using a lot of significant figures.

A measurement is accurate if you are able to give an answer close to the true value.

For example if a true value is 4.02 units a value of **4.51 is precise but not accurate**, a value of **4 is accurate but not precise. 4.01 is accurate and precise!**

When there is a limited range of alternative answers to a question (e.g. bigger/smaller, yes/no, true/false etc.) don't think that guessing is going to give you a 50% chance of scoring the mark. It is your reasoning that will determine success. **Examiners will usually expect you to support your choice of answer with the correct reason.**

Key points to remember

Materials behave differently when subjected to deforming forces. **Hooke's law** applies to materials that are not stretched too much. So, up to their limit of proportionality, the extension is directly proportional to the load applied.

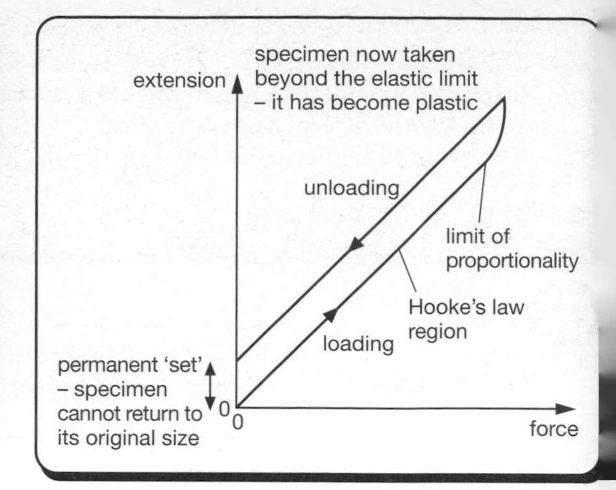

- When a material is stretched, part of it may be unable to return to its original size. When this happens, the material is said to have passed its **elastic limit** and to have become **plastic**.
- The elastic limit corresponds to an extension greater than the limit of the proportionality.

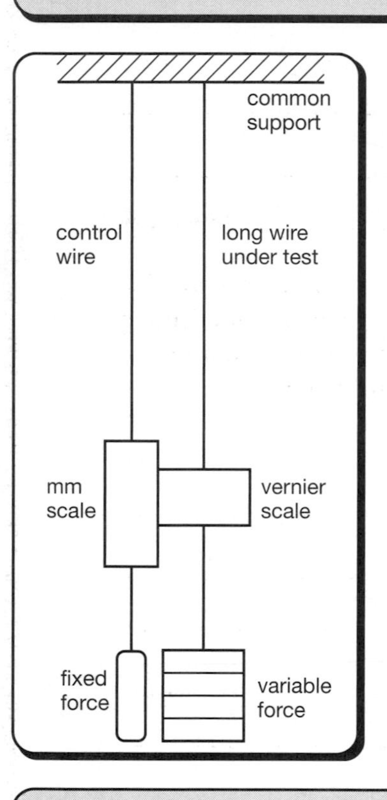

- $E = \frac{\sigma}{\varepsilon}$
- E = the Young modulus in Pa
- σ = stress (force/area) in Pa
- ε = strain (extension/original length) – no units

- When a wire is stretched, inter-atomic forces increase and it stores potential energy.
- The energy stored in a stretched wire is the area under a force extension graph and is given by

$$E_E = \frac{1}{2}Fx \text{ or } \frac{1}{2}kx^2$$

- Forces need to be large to stretch metal wires appreciably. Even then, the extension so small that you would need a vernier scale to measure it.
- For the linear part of a load against extension graph, the gradient should be multiplied by the original length of the wire and divided by the cross-sectional area to obtain a value for E.
- The diameter of the wire should be measured with a micrometer screw gauge and the area calculated from $\frac{1}{4}\pi d^2$.

Terms used to describe materials:

- **Yield stress** = value of stress for which the strain increases significantly for minimal increase in stress.
- **Fatigue** – situation in which materials fracture after undergoing loading/unloading cycles at values below normal breaking stress.
- **Creep** = a gradual increase in strain when a material has been under stress for a long time.
- **Ductile** describes a material that can be easily drawn into wire.
- **Brittle** describes a material that breaks without going through the plastic phase.

Types of material:

Crystalline Particles are arranged in a regular pattern. This can be in a single crystal e.g. diamond, or in grains e.g. polycrystalline metals, such as metals galvanised with zinc.

particles

Amorphous Particles have no long term order; they are arranged randomly e.g. glass, tar and sealing wax.

particles

Polymers Consist of long chains of molecules that tangle easily. When the chains are straightened, they can take strains of several hundred per cent.

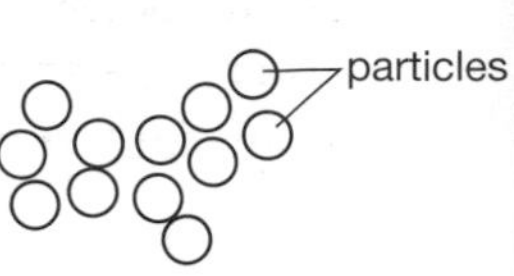

Question to try

Examiner's hints

- In **part (a)** don't ignore the word tensile in your definition.
- In **part (b)** don't forget to include the procedure you would use to ensure that the measurements are accurate.
- Remember, in part (c), that the material is ductile. Think, therefore, how the strain will change with increased stress.

Q1

(a) The Young modulus is defined as the ratio of *tensile stress* to *tensile strain*.

Explain what is meant by each of the terms in italics.

tensile stress ....

....

tensile strain ....

.... [3 marks]

(b) A long wire is suspended vertically and a load of 10 N is attached to its lower end. The extension of the wire is measured accurately. In order to obtain a value for the Young modulus of the material of the wire, two more quantities must be measured. State what these are and in each case indicate how an accurate measurement might be made.

quantity 1 ....

....

method of measurement ....

....

quantity 2 ....

....

method of measurement ....

.... [4 marks]

(c) Sketch a graph showing how stress and strain are related for a ductile substance and label important features.

stress

strain

[2 marks]

The answer to this question is on page 87.

[Total 9 marks]

Exam question and student's answer

(a) Some electrical components may be described as *non-ohmic*.

(i) Name an example, other than a diode, of a non-ohmic electrical component.

bulb ✗

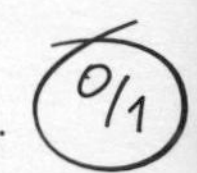

(ii) State how the current-voltage characteristic of your chosen components shows that it is non-ohmic.

its resistance changes with current ✓

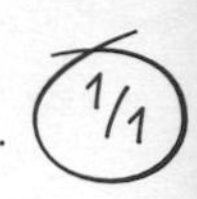

[2 marks]

(b) A semiconducting diode has special electrical properties that make it useful as an electrical component.

(i) Sketch on the grid the current-voltage characteristic of a diode.

Fig. 6.1

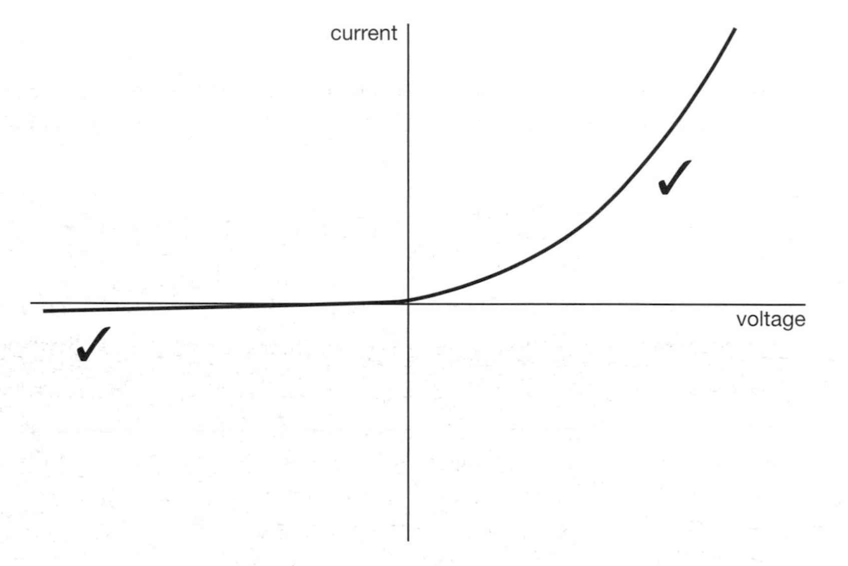

(ii) State, with reference to the current-voltage characteristic you have drawn, how the resistance of the diode varies with the potential difference across its terminals for reverse bias and for forward bias.

reverse biased: it doesn't allow any current to go through ✗

forward biased: its resistance increases proportionally with time ✗

[4 marks]

[Total 6 marks]

How to score full marks

Part (a)(i)

This answer is not correct. It is true that a bulb filament gives a non-linear characteristic **but this is because the temperature of the filament changes with current**. Since the temperature is not being held constant, Ohm's law is invalidated. If the temperature were to be held constant then Ohm's law would apply.

Part (a)(ii)

This part only just gets the mark in terms of "error carried forwards". This is because a component that was non-ohmic would show a variation in resistance (it would be non-linear) – **this is not fully spelled out in the answer**. The comment for part (a)(i) still applies. **A better answer would be to say that the graph is "non-ohmic" because it is a curve**.

Part (b)(i)

There are two marks for this characteristic and the student has gained them both by showing a curved forward bias characteristic and zero current for reverse bias. However, **the graph is not perfect as there should be zero current on forward bias until about 0.7 V when it should start to increase as a curve**.

(*Forward bias* is when the positive side of the power supply is connected to the positive side of the diode; when the positive side of the power supply is connected to the negative side, this is *reverse bias*).

Part (b)(ii)

This answer for *reversed bias* is not wrong **but it does not answer the question**. The student needs to say that **the resistance does not change** (it is very high, or infinity in an ideal diode).

For forward bias the student is now answering correctly, but the answer is wrong! **The resistance falls as the potential difference increases**.

Don't make these mistakes ...

Don't give vague, general answers in the hope that your lack of knowledge will not be caught out. **Imprecise answers will not normally be given the benefit of the doubt**. The student's answer bulb, in the question, is imprecise because a bulb or lamp is made up of a number of parts and it is not clear to which part the answer refers (case, holder, filament, gas inside etc.).

Avoid saying that quantities simply "change" with changing conditions. This is not a complete answer and begs the question, 'In what way?'. Properties **increase** or **decrease** with variation in conditions and so **you should say that the resistance (of the diode) decreases as the potential difference increases**.

Key points to remember

- **Current** = charge flowing in a fixed time: $I = \frac{\Delta Q}{\Delta t}$
- **Units**: $A = Cs^{-1}$
- The current is the gradient of a graph of charge against time.
- When charge varies in a non-linear way you need to find the gradient of the tangent to the curve at a chosen time.
- The charge passing in a given time is the area under a graph of current against time.

Think of circuits as if there were no meters in them. Meters should not change the circuit at all and so ammeters must have a very low resistance and voltmeters must have a very high resistance (each when compared with components).

Fig. 6.2

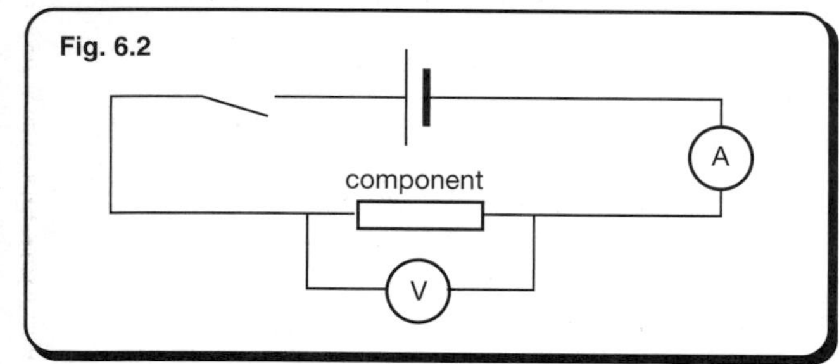

- **Potential difference** (p.d.) = the energy per coulomb that is transferred from electrical to other forms: $V = \frac{W}{Q}$
- **Units**: $V = JC^{-1}$
- **Resistance** (of a circuit component) = p.d.across the component divided by the current in it:

$$R = \frac{V}{I}$$

- **Units**: $\Omega = VA^{-1}$
- **Ohm's law** applies only to a conductor held at constant temperature and states that the resistance will be constant.

- You must know the circuit shown in Fig 6.2. The resistance of the component can be calculated by dividing the voltmeter reading by the ammeter reading.
- To obtain a range of readings, alter the current by including a variable resistor in series with the ammeter.
- The order of the components in the series part of the circuit does not matter since there is the same current throughout the circuit.

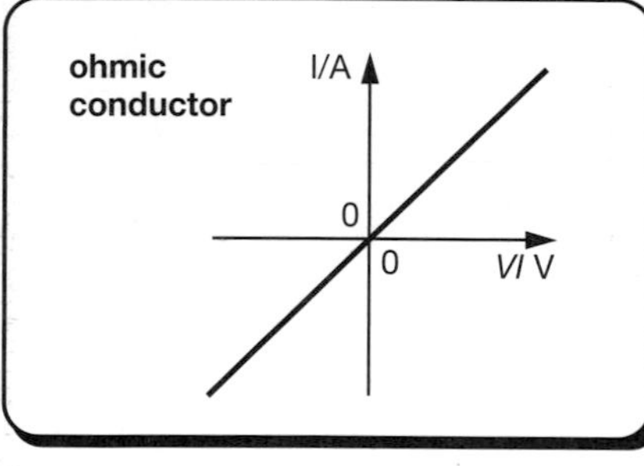

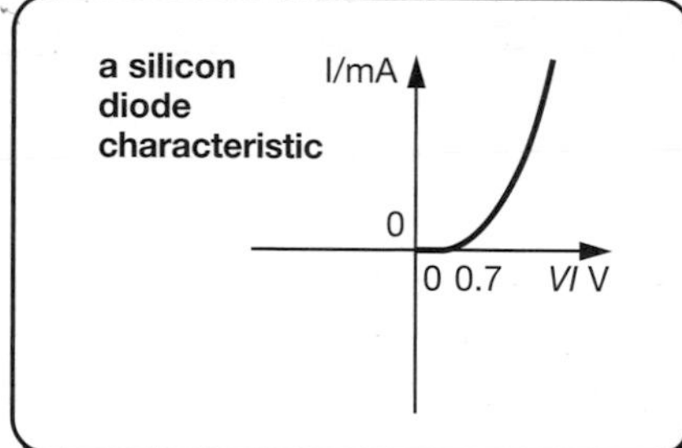

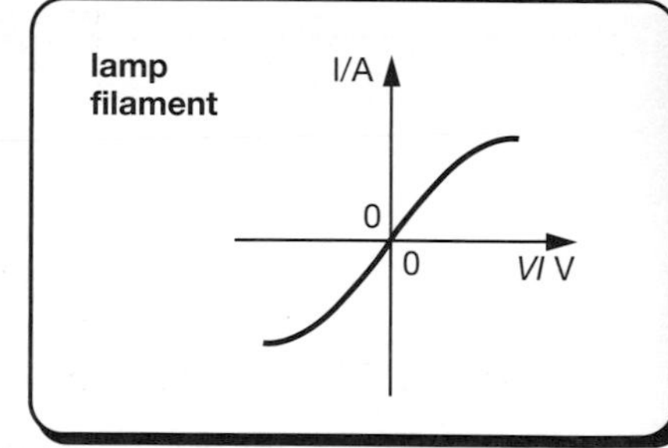

- **Resistivity** (ρ) is a property of a material: $\rho = \frac{RA}{l}$
- **Units**: Ωm (**NOT** Ωm^{-1})
- **Conductivity** (σ) = $1/\rho$
- To measure resistivity, a wire is placed in the circuit in Fig. 6.2 as the component, and the resistance of different lengths is then found.
- A graph of resistance against length gives a straight line of gradient resistance per unit length.

- The current, I, in a wire can be shown to be given by the equation $I = neAv$
- Here n = number of charge carriers per unit volume, e = charge on the charge carrier, A = cross-sectional area of the wire and v = mean drift velocity of the charge carriers.
- Charge carriers in metal wires are electrons.

Questions to try

Examiner's hints for question 1
Part (a) When the question asks for words, don't write a symbol equation.
Part (b) Do write down the equation to focus your answer.

Q1

(a) (i) Give, in words, the equation which is used to define charge.

....

(ii) State the SI unit of charge.

....

(iii) Define potential difference and its unit, the volt.

potential difference ....

....

volt ....

....

[5 marks]

(b) In the circuit of Fig 6.3, the 6.0 V d.c. supply has negligible internal resistance.

Fig. 6.3

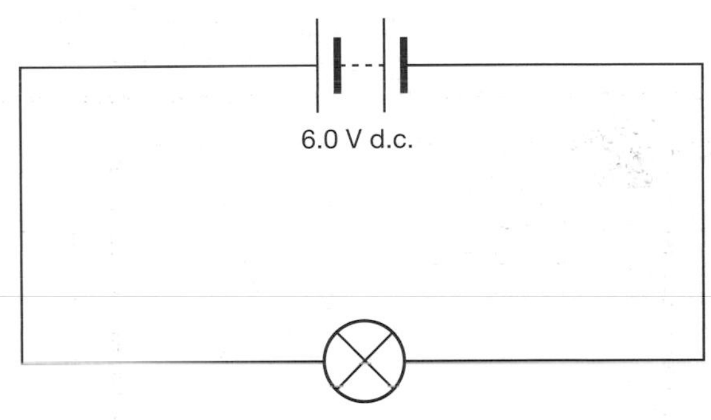

(i) On Fig 6.3, show, by means of arrows

1. the convential current in the circuit (label this arrow C)

2. the electron flow in the circuit (label this arrow E).

[2 marks]

(ii) Calculate the energy transfer in the bulb when a charge of 15 C passes through it.

energy transfer = ...J

[2 marks]

[Total 9 marks]

Examiner's hints for question 2

Part (a) This question asks you to *show that* ... It is essential that you show all your working here. You cannot afford to fudge an answer to this type of question.

Part (b) You cannot get full marks for this type of question unless you address each of the bullet points.

Q2

(a) A pencil "lead" is made from non-metallic material which has a resistivity, at room temperature, of 4.0×10^{-3} Ωm. A piece of this material has a length of 20 mm and a diameter of 1.40 mm.

Show that the resistance of this specimen, to two significant figures, is 52 Ω.

....

....

....

[2 marks]

(b) Given a specimen of the pencil "lead" described in part (a) with similar dimensions, describe an experiment you could carry out in the school or college laboratory to verify that the resistivity of the material is equal to the value quoted in part (a).

Your description should include

- a labelled circuit diagram,
- details of the measurements you would make,
- an account of how you would use your measurements to determine the result.

....

....

....

....

....

....

[8 marks]

(c) During an experiment such as that described in part (b), a specimen of pencil "lead" is found to have a resistance of 52 Ω when the current through it is 250 mA.

Calculate the power dissipated in the specimen under these conditions.

....

....

....

[2 marks]

The answers to these questions are on page 88.

[Total 12 marks]

Exam question and student's answer

This question is about a potential divider circuit.

The circuit diagram shows a 12 V battery connected across a variable resistor and a 300 Ω resistor. When a high resistance voltmeter is connected across the 300 Ω resistor the p.d. registered by the voltmeter changes as the resistance of the variable resistor, R, changes.

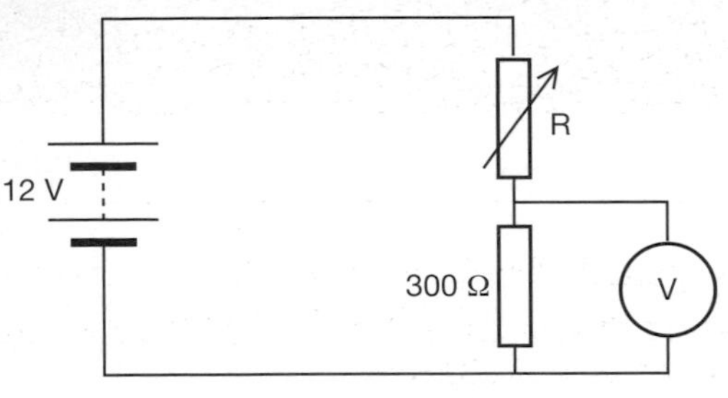

Fig. 7.1

(a) (i) Describe what happens to the voltmeter reading as R decreases.

R decreases. The voltmeter reading will increase ✓

(ii) Show that the p.d. V across the 300 Ω resistor is given by the expression: $V = \frac{3600}{300 + R}$.

$$12 \times \frac{300}{300 + R} = V \quad \therefore \quad \frac{3600}{300 + R} = V$$ ✓

[3 marks]

(b) The variable resistor is replaced by a thermistor.

The graph shows how the resistance of the thermistor varies with temperature.

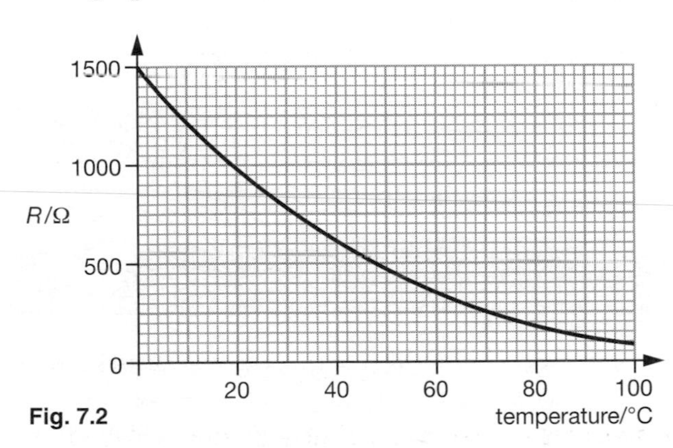

Fig. 7.2

(i) Use the graph to find the resistance R of the thermistor at 0 °C and 100 °C.

Resistance R of thermistor at 0 °C

= 1500 Ω ✓

Resistance R of thermistor at 100 °C

= 80 Ω ✓

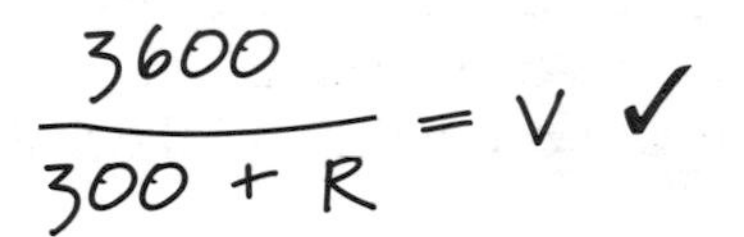

(ii) Show that the voltmeter readings at 0 °C and 100 °C are approximately 2 V and 10 V, respectively.

$$V = \frac{V_s R_1}{R_1 + R_2} \Rightarrow \frac{12 \times 300}{300 + 1500} = 2\text{V}$$ ✓ ✓

$$\Rightarrow \frac{12 \times 300}{300 + 80} = 9.5\text{V}$$ ✓

(iii) Some instruments have a linear response, others are non-linear. Using the graph, discuss the response of the thermistor.

graph is a curve so response is not linear ✓

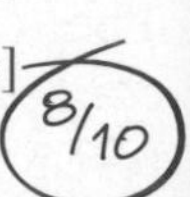

[7 marks]

[Total 10 marks]

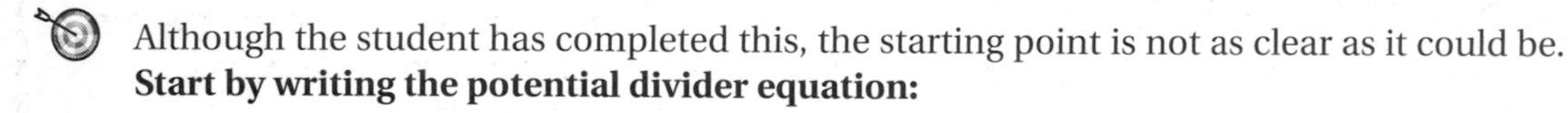

How to score full marks

Part (a)(i)

This is correct, as **the question requires only a description of what will happen.**

Part (a)(ii)

Although the student has completed this, the starting point is not as clear as it could be. **Start by writing the potential divider equation:**

$$V = \frac{V_s R_1}{R_1 + R_2}$$

Parts (b)(i) and (b)(ii)

In each case the student **has read the graph accurately** and done enough to show that the voltmeter readings are approximately 2 V and 10 V respectively.

Part (b)(iii)

The student explains why the response is non-linear and scores the first mark. **You need to discuss what that means** in order to gain the second mark available. You could write: "if the response were linear there would be equal drops in resistance for equal increases in temperature; here there are larger resistance drops for the lower temperature increases."

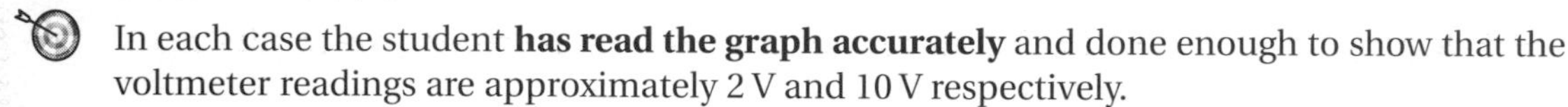

Don't make these mistakes ...

Don't assume that you will be given credit by starting a calculation or proof mid-way through it. **Always start by writing down the equation.**

Don't think that drawings and graphs on the exam paper are sacrosanct. In order to ensure that you make an accurate reading, mark in horizontal and vertical lines. The correct marks will also convince the examiner that you know what you are doing!

Don't ignore the word 'discuss'. It is a key word like 'explain' and it means that you are expected to do more than simply state fact. You should consider the options that could happen under various circumstances clearly and concisely. You will not gain extra marks for writing very long 'wordy' answers.

Key points to remember

- **Emf** is the energy change per coulomb when other forms of energy are converted into electrical energy.
- All power supplies have some resistance which will mean that some of the source energy is converted into thermal energy in the *internal resistance.*

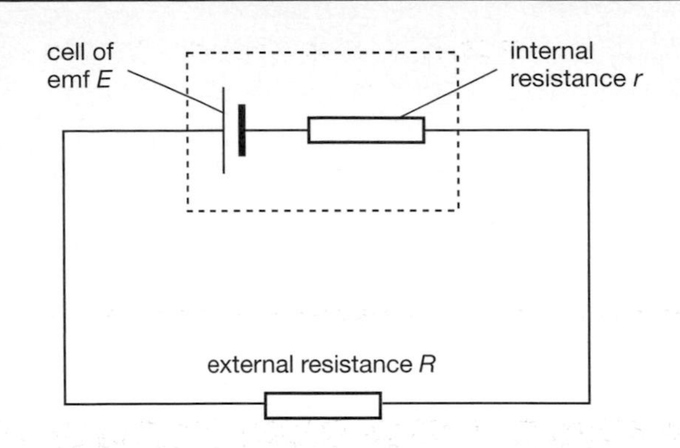

- The resistor in the broken-lined box is the internal resistance of the cell.
- $E = I(R + r)$

  $E = V + Ir$

  where V is the p.d. across the external resistor and Ir is often called the 'lost volts' which cannot be used externally.

- The power dissipated inside the cell is given by:

  $P_{int} = I^2 r$
- The power dissipated in the external resistor is given by:

  $P_{ext} = I^2 R$
- The maximum power will be dissipated externally when:

  $R = r$

Variable resistors can be used in circuits in two ways:

- As a method of **changing the current in the circuit** by increasing or decreasing the resistance.

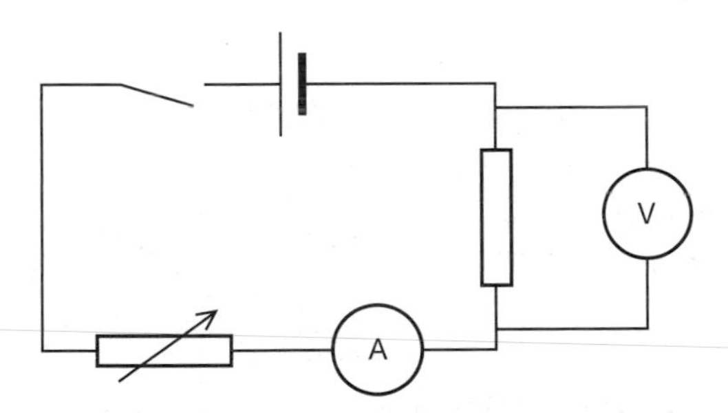

Increasing the resistance of the variable resistor **decreases** the current through the circuit – this is a simple way of investigating current/voltage relationships.

The maximum current for a particular power supply is limited by the internal resistance of the power supply and the resistance of the component being investigated.

- As a method of **changing the voltage supplied to a component** by sharing the voltage across the two portions of the variable resistance – i.e. as a potential divider.

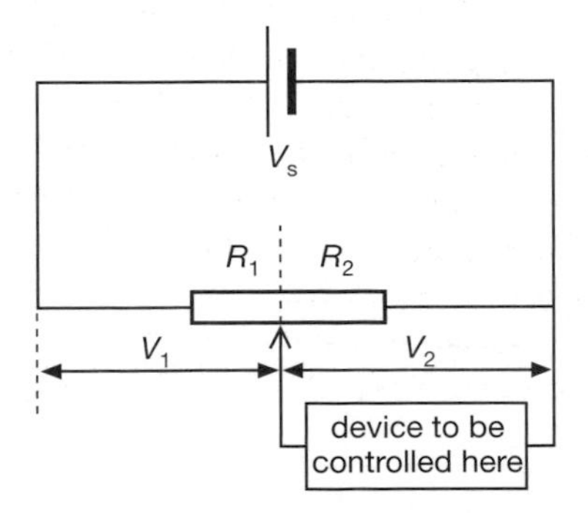

- With a supply voltage of negligible internal resistance the p.d.'s across the resistors R_1 and R_2 will be given by:

  $V_1 = V_s \times \frac{R_1}{R_1 + R_2}$ and $V_2 = V_s \times \frac{R_2}{R_1 + R_2}$

  respectively.
- This assumes that the resistance of the device being controlled is much greater than R_1 and R_2.
- The maximum voltage available to the device is V_s.

Questions to try

Examiner's hints for question 1

- This question tests your understanding of the different uses of a variable resistor. It can be used as a rheostat and a potential divider. You need to be clear about the advantages and disadvantages of each. You also need to consider how to calculate the current when you know the voltage and the resistance **(part (b)(i))** and when you know the power and the voltage **(part (b)(ii))**.

(a) The current flowing through a torch bulb can be controlled by a variable resistor using either of the two circuit arrangements shown. Fig 7.3 is called a potential divider arrangement and Fig 7.4 may be called a rheostat arrangement. For each of these two methods explain **one** advantage and **one** disadvantage.

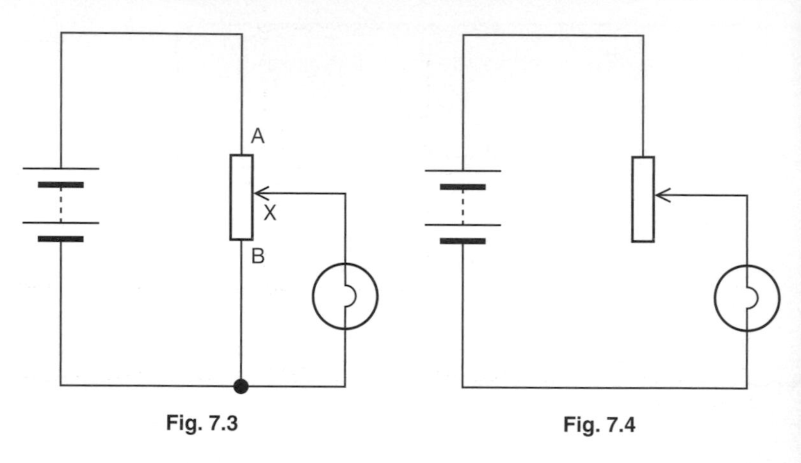

Fig. 7.3 Fig. 7.4

potential divider

advantage ....

....

disadvantage ....

....

rheostat

advantage ....

....

disadvantage ....

.... [4 marks]

(b) In Fig 7.4, the variable resistor has a total resistance of 16Ω. When the slider of the variable resistor is set at X, exactly mid-way along AB, the bulb works according to its specification of 2.0V, 500mW. Calculate

(i) the current through section XB of the variable resistance,

....

(ii) the current through section AX of the variable resistance.

.... [2 marks]

[Total 6 marks]

Examiner's hints for question 2

- It is essential that you know your definitions of e.m.f. and internal resistance verbatim. Practise writing them out and check against a book definition.
- In **part (b)** the voltmeter is measuring the external p.d. across the cell (which is the same as that across the resistor)

Q2

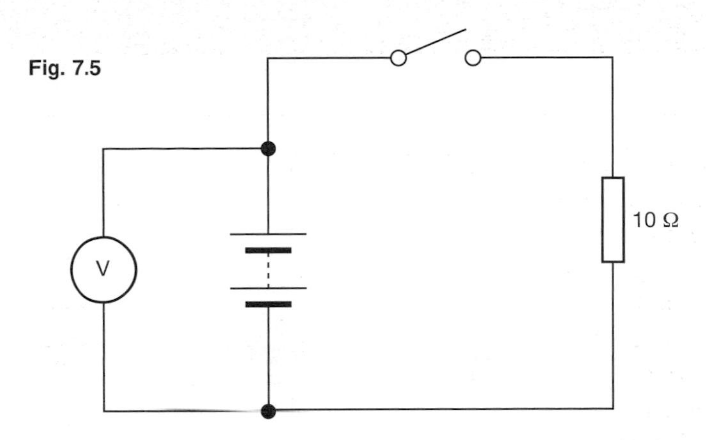

A battery is connected to a 10 Ω resistor as shown. The e.m.f. (electromotive force) of the battery is 12 V.

(a) (i) Explain what is meant by the e.m.f. of a battery.

....

....

....

....

(ii) When the switch is open the voltmeter reads 12.0 V and when it is closed it reads 11.5 V. Explain why the readings are different.

....

....

....

[3 marks]

(b) Calculate the internal resistance of the battery.

....

....

....

[3 marks]

The answers to these questions are on pages 88 and 90. [Total 6 marks]

Exam question and student's answer

1 In each of the following circuits the battery has negligible internal resistance and the bulbs are identical.

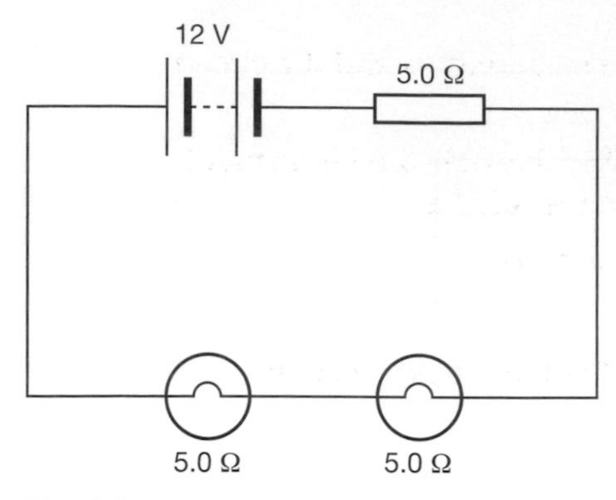

Fig. 8.1

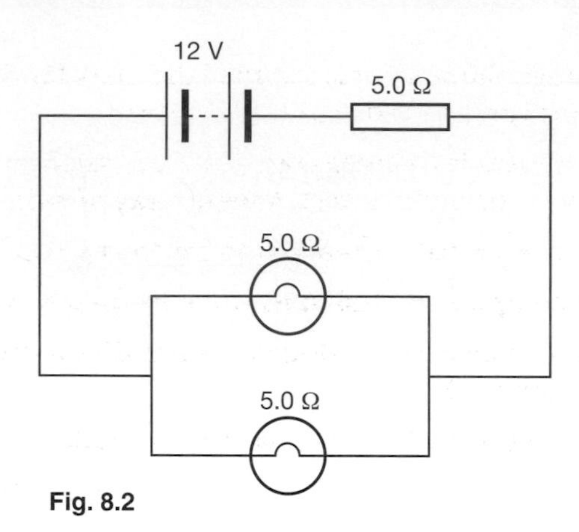

Fig. 8.2

(a) For the circuit shown in Fig 8.1 calculate

(i) the current flowing through each bulb,

$I = \frac{V}{R} = \frac{12}{15} = 0.8A$ ✓ (1/1)

(ii) the power dissipated in each bulb.

$P = IV$

$= 0.8 \times 12 = \frac{9.6}{3} = 3.2W$ ✓ (1/1)

[2 marks]

(b) In the circuit shown in Fig 8.2 calculate the current flowing through each bulb.

$1/R = 1/5 + 1/5 \quad \therefore \quad R = 2.5\ \Omega$

$I = \frac{12}{2.5 + 5}$ ✓ $= 1.6$ A ✓ ^

[3 marks] (2/3)

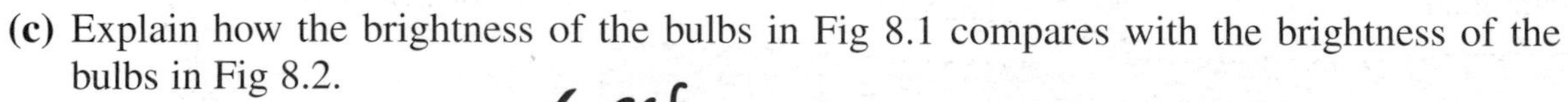

(c) Explain how the brightness of the bulbs in Fig 8.1 compares with the brightness of the bulbs in Fig 8.2.

✓ ecf

The bulbs are brighter in Fig 8.2 as they are in parallel so there is more current in each bulb – dim in Fig 8.1 – series. ✗

(1/2)

[2 marks]

[Total 7 marks]

(5/7)

How to score full marks

Part (a) (i)

This part is clearly answered. The student has realised that the resistance of the resistor must be added to that of the bulbs in series.

Part (a) (ii)

Although the student is awarded the mark for this (and appears to understand the circuit) the answer is flawed: $0.8\text{A} \times 12\text{ V} \neq 9.6/3$

The student is trying to cram two expressions into one and has derived something that is mathematically incorrect. **A better way of writing this answer would be:**

$0.8\text{A} \times 12\text{ V} = 9.6\text{W}$ (total power for the two bulbs and the resistor)

each component dissipates 1/3 of the power = 3.2W.

An alternative way to find the power dissipated in each bulb filament is to use the equation $P = I^2R$

Thus power dissipated in each bulb = $(0.8\text{A})^2 \times 5\Omega$

$= 3.2\text{W}$

Part (b)

The student has found the current flowing through the main branch of the circuit **but hasn't realised that this splits into two equal parts as it flows in the bulbs**. Thus the final answer should be 0.8A.

Part (c)

The first part of this answer is correct according to the student's previous answer and so is awarded an error carried forward mark (ecf). However, this statement is impossible to justify and so the second mark can't be scored here.

The correct answer would be that all the bulbs would be equally bright in the two circuits since they all carry the same current (0.8A).

Don't make these mistakes ...

Avoid writing relationships that are mathematically impossible. In this case the student was fortunate because there was only one mark available. Had there also been a mark awarded for showing working, then this student would have lost this "working" mark.

Remember that in **parallel circuits the voltage is the same across each of the parallel components**. The current in each of the components is equal to the voltage divided by the resistance of the component.

Errors from previously calculated values are carried forward to subsequent parts of a question. **Even when you have made an early mistake all is not lost – so carry on with your answer**. As in this student's answer, you may end up trying to argue a point that you know cannot be true. If you find yourself in this situation don't write what is obviously rubbish – return to the whole question and rework it after you have completed the remainder of the questions.

Key points to remember

Although circuits may seem complex they can usually be simplified into a single resistor with a p.d. across it and a current going through it. This is called an **equivalent circuit**.

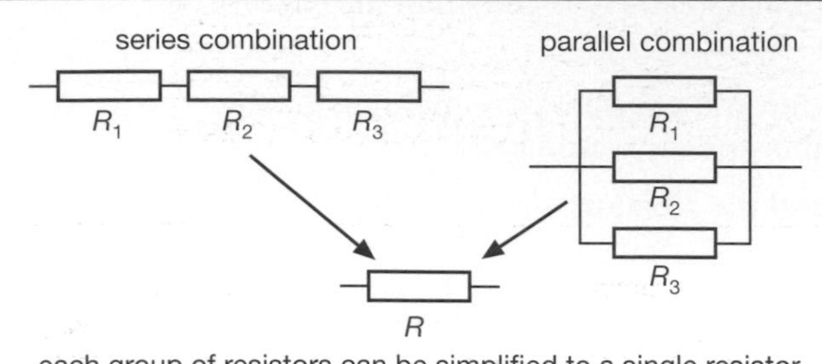

each group of resistors can be simplified to a single resistor

series: $R = R_1 + R_2 + R_3$

parallel: $\frac{1}{R} = \frac{1}{R_1} + \frac{1}{R_2} + \frac{1}{R_3}$

- A second way of writing the parallel formula for just two resistors is:

$$R = \frac{R_1 \times R_2}{R_1 + R_2}$$

- This is probably easier to use than the reciprocal formula but it can only be used (in this form) for two resistors.
- If you choose to use the reciprocal formula – don't forget to invert the equation after you have added the two reciprocals together.
- The equivalent resistance of two resistors in parallel will always be smaller than either of the resistors.

The laws of conservation of charge and energy apply to a circuit. These can be written as **Kirchhoff's laws** or **rules**:

- At a junction in a circuit the total current leaving the junction equals the total current entering the junction or algebraically $\Sigma I = 0$ (which means $I_1 + I_2 = I_3 + I_4 + I_5$)

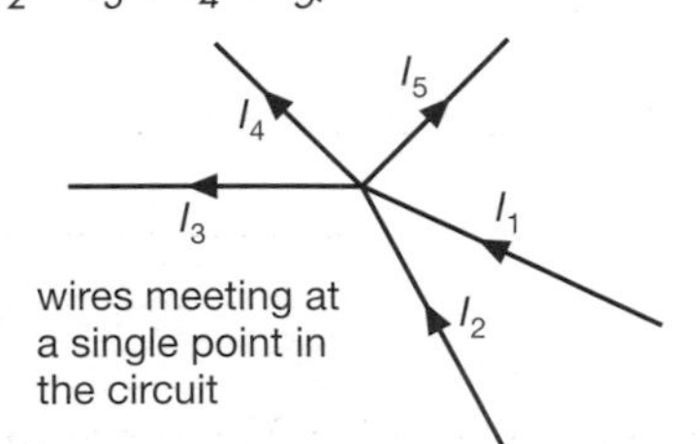

- For any closed loop in a circuit the sum of the e.m.f's is equal to the sum of the p.d.'s or algebraically $\Sigma E = \Sigma I R$

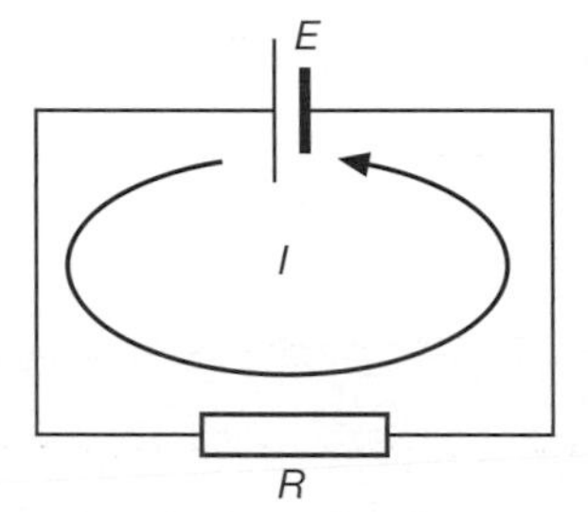

For this circuit with just one cell and one resistor this law gives

$E = IR$

This should be familiar to you!

Application of Kirchhoff's laws

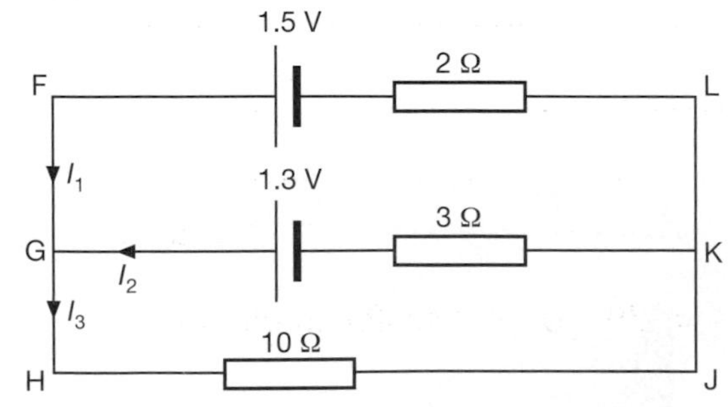

When we wish to calculate the p.d. across the 10 Ω resistor we can apply Kirchhoff's laws:

The first law tells us: $I_3 = I_1 + I_2$ .... (i)

Applying the second law to the loop FHJL:
$1.5\text{V} = I_3 \times 10\Omega + I_1 \times 2\Omega$ ....(ii)

and to the loop GHJK:
$1.3\text{V} = I_3 \times 10\Omega + I_2 + 3\Omega +$ ....(iii)

The physics stops at equation (iii) and the rest is algebra – however the answers are provided at the back of the book if you want to practise your simultaneous equations (see page 89). BUT having calculated the values for the currents you do need to be able to interpret the meaning of your values.

- A **negative current** simply means that your original arrow for that current was in the wrong direction.
- The **value of the current** multiplied by the **resistance** that it passes through indicates the **p.d.** across that resistor.

Question to try

Examiner's hints

- The horizontal lines are the heating elements. The shaded bands are the contacts. You must decide the configuration in which to connect each set of heating elements. You need not go through the whole calculation twice.
- Write down the power equation for a resistor. Think what effect halving the voltage has on the current. Does the resistance change from its normal working value? In the light of this, decide on the overall effect on the power.

Q1

Fig. 8.3 shows two methods of connecting eight heating elements which make up a car rear window heater. The heater is connected to a 12V car battery. Each element used in circuit P has a resistance of 24 Ω; each used in circuit S has a resistance of 0.50 Ω.

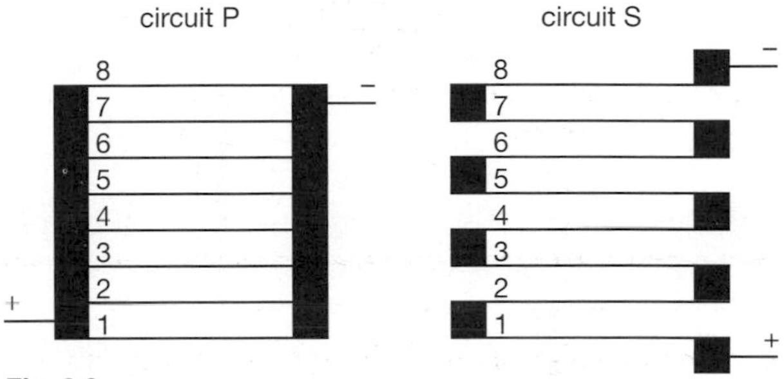

circuit P

8 7 6 5 4 3 2 1 + −

circuit S

8 7 6 5 4 3 2 1 − +

Fig. 8.3

(a) Calculate the current drawn from the battery for each circuit. Show your working.

Circuit P ....

....

Circuit S ....

....

[5 marks]

(b) Elements 3 and 4 burn out in each circuit and no longer conduct electricity. What are the new values of the currents in each circuit?

Circuit P ....

....

Circuit S ....

....

[2 marks]

(c) What effect would halving the battery voltage have on the power transfer in circuit P? Explain your answer.

....

....

....

[2 marks]

The answer to this question is on page 89.

[Total 9 marks]

Exam question and student's answer

1 **Fig. 9.1** shows the variation of displacement with distance from the source for a longitudinal mechanical wave at a particular time t.

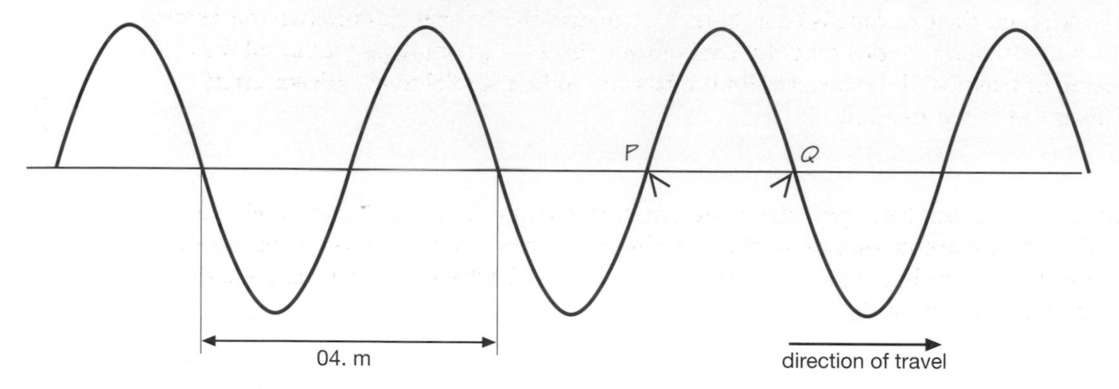

(a) Describe briefly how the propagation of a mechanical longitudinal wave is different from a mechanical transverse wave.

Transverse waves are those that oscillate across the direction of travel, longitudinal waves oscillate along the direction of travel.

[3 marks]

(b) Indicate with letters P and Q on Fig. 9.1, two points that are oscillating with a phase difference of 180°.

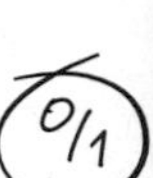

[1 mark]

(c) The speed of the wave is $20\,\mathrm{m\,s^{-1}}$. Show on Fig. 9.1 how the displacement varies with the distance from the source 5.0 ms later than at the time t shown on the original graph. Do any calculations needed in the space below.

$$\frac{20}{0.4} = 50\,\text{Hz}$$

$$50 \times \frac{5}{1000} = 0.25 \text{ of an oscillation moved in 5 ms}$$

[2 marks]

[Total 6 marks]

How to score full marks

Part (a)

The student's answer makes no reference to the 'mechanical' aspect of the question. **You have to say that the particles in the medium must oscillate for a mechanical wave to propagate**. It is unusual to refer to 'across' and 'along' the direction of travel – although the meaning is clear here. The answer has been given the two marks for this. You would do better to describe the **transverse vibration as being perpendicular to the direction of travel and the longitudinal vibration to be parallel to the direction of travel**. This is then unambiguous.

Part (b)

The student's answer has been penalised here. Although **P** and **Q** appear to be roughly in the right place, **there are no points marked on the curve**. It is impossible to be absolutely sure that the student really did mean the points near **P** and **Q** where the curve crosses the axis (shown by the insertion signs).

Part (c)

The student seems unsure of what to do here and has simply divided the wave speed by the wavelength to calculate the frequency (as correctly shown by the units of Hz). He has then multiplied the frequency by the time delay – although this has been calculated to be 1/4 of an oscillation, **this is rather vague and not quite good enough to gain the mark available**. He has not attempted to mark in the position of the wave at this time.

You should calculate the distance that the wave has moved in 5 ms by multiplying the wave speed by the time:

distance moved = $20\,\text{ms}^{-1} \times 5 \times 10^{-3}\,\text{s} = 0.1\,\text{m}$

this distance is equivalent to a **quarter of a wavelength** as shown below:

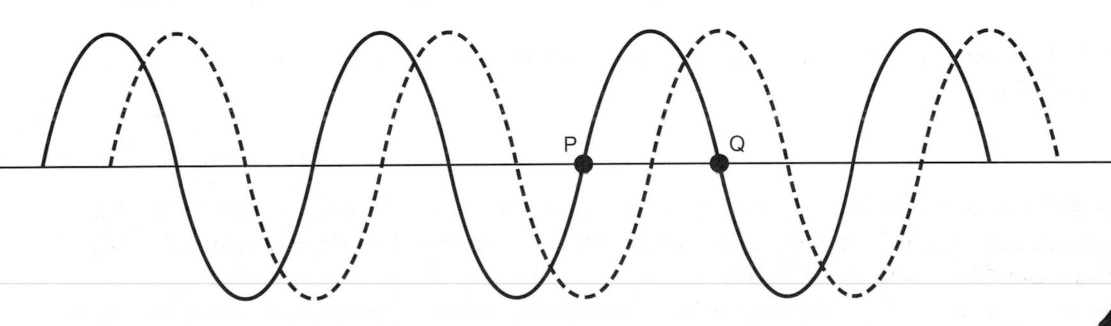

Don't make these mistakes ...

Don't ignore parts of phrases in the question as being unimportant. The examiners have intended you to take notices of the word 'mechanical' – that's why they put it in.

Do answer the question asked. You need to mark in points if that is what the question tells you – if you leave things vague, the examiner will assume you don't know what you are doing. Be precise.

Do have a go at completing the full question. In this case the student calculated the fraction of the **period** that turned out to be consistent with the question (even though it did require a fraction of the **wavelength**). Drawing the $\frac{1}{4}$ period lag could have gained the student both marks since it would have clarified the meaning of the student's calculation.

Be careful with wave diagrams – look at the quantity on the horizontal axis. Crest to crest on a time axis is a **period** but on a distance axis it is a **wavelength**.

Key points to remember

- **Waves** transfer energy from place to place.
- In **longitudinal waves**, the energy travels in a direction parallel to the direction of oscillation.
- In **transverse waves**, the energy travels in a direction at right angles to the direction of oscillation.
- **Mechanical waves** always need a medium in order to be able to travel.
- **Electromagnetic waves** require no medium and can travel through a vacuum.

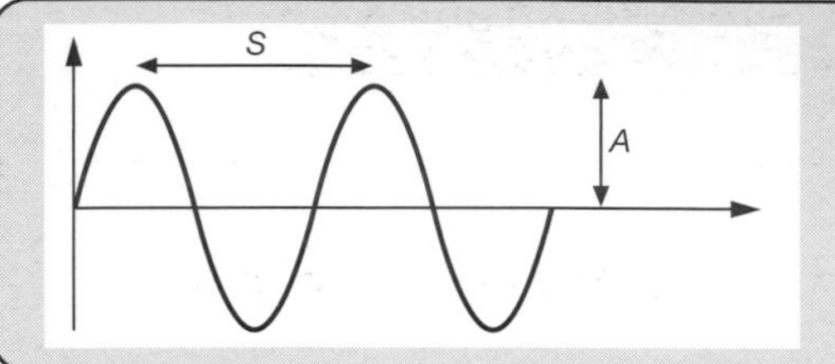

- In the diagram on the left, when the horizontal axis is time, S represents the period of the wave (T).
- When the horizontal axis is distance then S represents the wavelength of the wave (λ).
- A is the amplitude or maximum displacement of the wave.

- All waves reflect, refract, diffract, and interfere. Only transverse waves can be polarised.

- $v = f\lambda$ v = wave speed in ms^{-1}; f = frequency in Hz
- $T = 1/f$

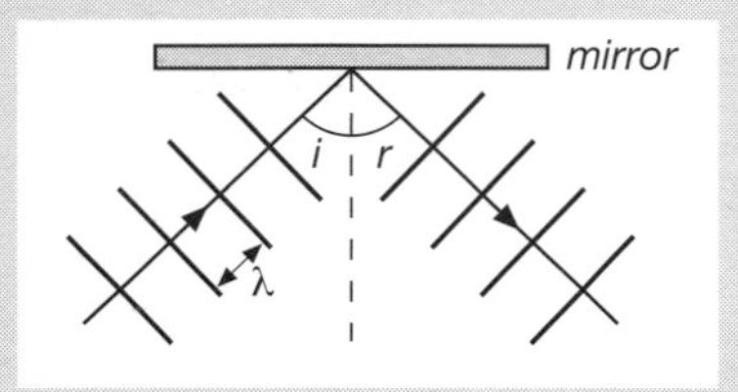

- No change in wavelength.
- No change in speed.
- Angles of incidence and reflection are equal.

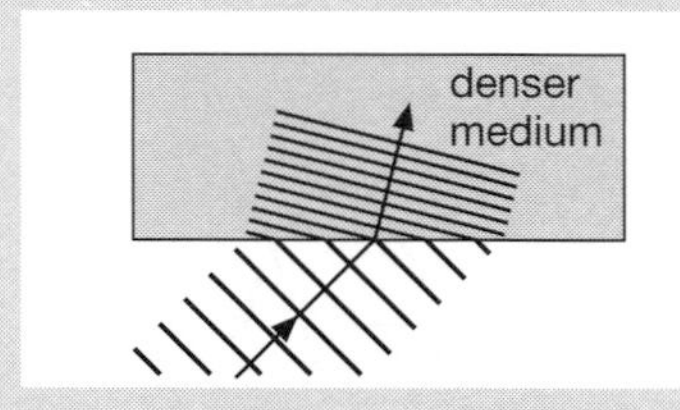

- Speed and wavelength decrease as wave enters denser medium.
- Frequency stays constant.

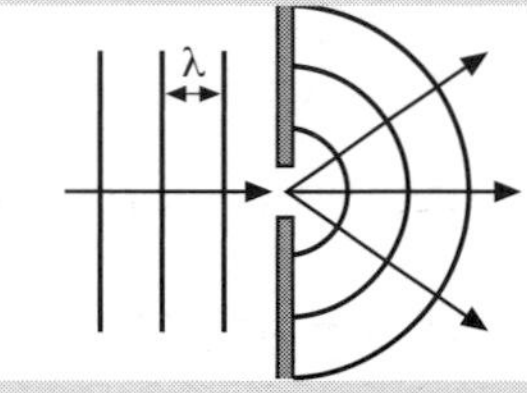

- Significant diffraction only seen when the slit width is approximately the size of the wavelength.

- Two coherent sets of waves can be made to interfere consistently, e.g. with a pair of dippers in a ripple tank.

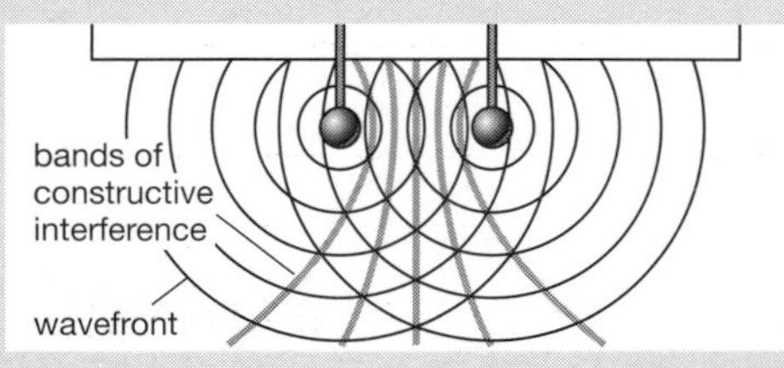

- This produces a series of evenly-spaced fringes of alternating high and low intensity.
- Coherent waves have the same frequency and a constant phase relationship.

Superposition of waves:
Where two waves meet, the total displacement is the vector sum of the displacements of the individual waves.

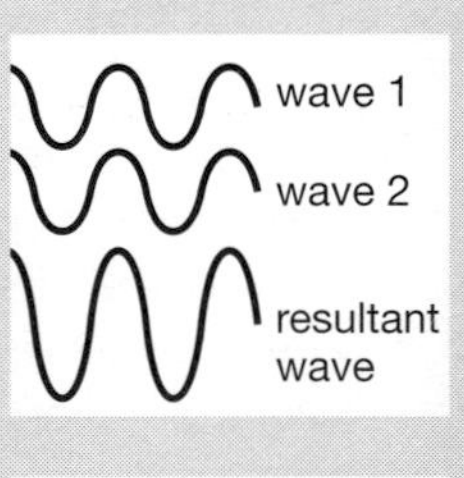

Constructive interference:
When waves meet in phase their displacements add together making a resultant wave of amplitude, which equals the sum of amplitudes of the component waves.

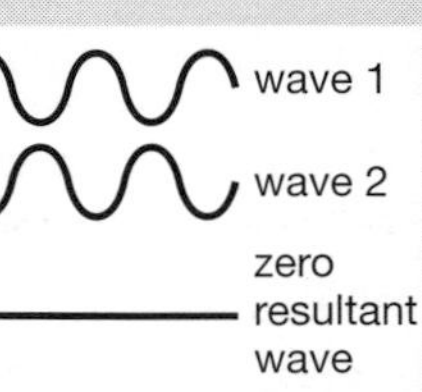

Destructive interference:
When waves of equal amplitudes meet antiphase they cancel each other out completely giving zero resultant.

Questions to try

Examiner's hints for question 1

- This is largely a descriptive question and tests your use of the language of physics. In (a) you need to make sure that you consider the relationship between adjacent particles and use ideas including amplitude, frequency and phase in your answer. Remember four marks means four points. Ignore the word 'progressive' – it is used here to distinguish the wave from a standing wave (see Chapter 10). In (b) you need to be able to give exact definitions of these quantities. Vague answers will not score marks.
- Part (c) uses the wave equation: $v = f\lambda$.

Q1

(a) Describe what happens to the particles of a medium when a progressive wave passes through it.

....

....

....

....

[4 marks]

(b) Explain what is meant by:

(i) the *amplitude* of a wave

[1 mark]

(ii) the *frequency* of a wave

[1 mark]

(iii) the *wavelength* of a wave

[2 marks]

(c) A wave has a wavelength of 3.0 m and a frequency of 5.0 Hz. Calculate the wave-velocity.

....

....

[2 marks]

[Total 10 marks]

Examiner's hint for question 2

- This is a context question and it is important to focus on the physics of the situation in the first part. The second part goes on to develop the context of the application and is meant to force you to apply some ideas that are used at GCSE. Do not be distracted by the context of such questions.

Fig. 9.3 shows one way of transmitting information from a transmitter to a receiver that is not directly in its line of sight.

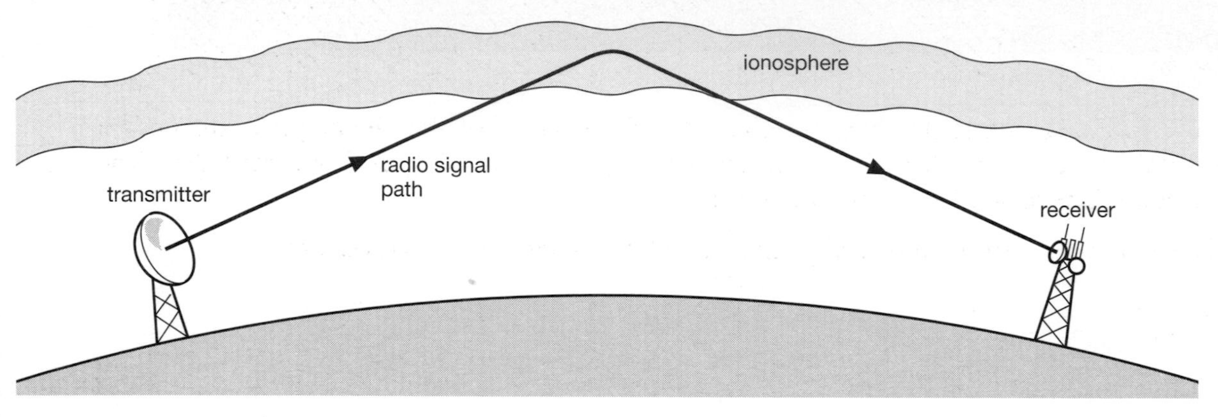

Fig. 9.3

(a) State the property of waves that enables the signals to be transmitted back to Earth by the ionosphere.

....

....

[1 mark]

(b) (i) Describe briefly one other method of transmitting signals to a receiver that is not in the line of sight of a transmitter.

....

....

[2 marks]

(ii) State one advantage or disadvantage of the method you have described when compared with that in **Fig. 9.3**.

....

....

[1 mark]

[Total 4 marks]

The answers to these questions are on page 89.

Exam question and student's answer

1 A student performs an experiment to investigate how the speed of the waves on a stretched string is affected by the tension T. The tension T is equal to the weight of the mass hanging over a pulley as shown in **Fig 10.1**:

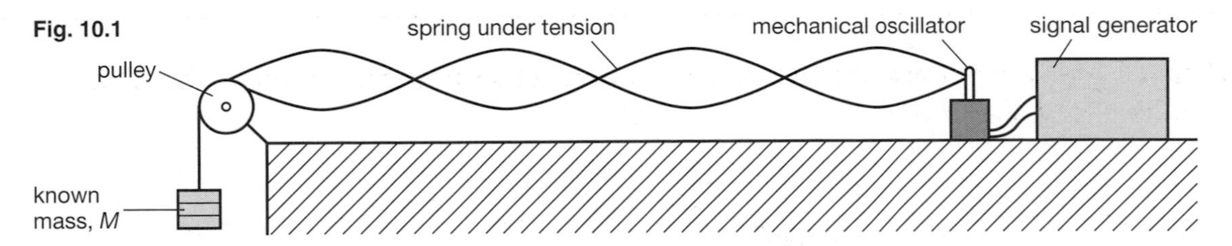

Fig. 10.1

For a particular mass, the student adjusts the frequency f of the signal generator until a standing wave is established. The frequency f and the wavelength λ are recorded. The procedure is repeated for a rage of masses.

The student finds from an A level textbook that n is given by the expression:

$$v = \sqrt{\frac{T}{\mu}}$$

Where μ is the mass per unit length of the string.

The student decides to see if the data support this expression. She starts to process her data and plot it on a graph.

Use the information in the table to add two more points to the graph. Record the results of your calculations in the table.

[3 marks]

M/kg	f/Hz	λ/m	M/kg	f/Hz	λ/m
0.16	30.6	0.37	0.24	29.4	0.44
0.20	30.0	0.41	0.28	28.8	0.48

✗ 0/3

Draw the line of best fit through the points on the graph. ✗

[1 mark]

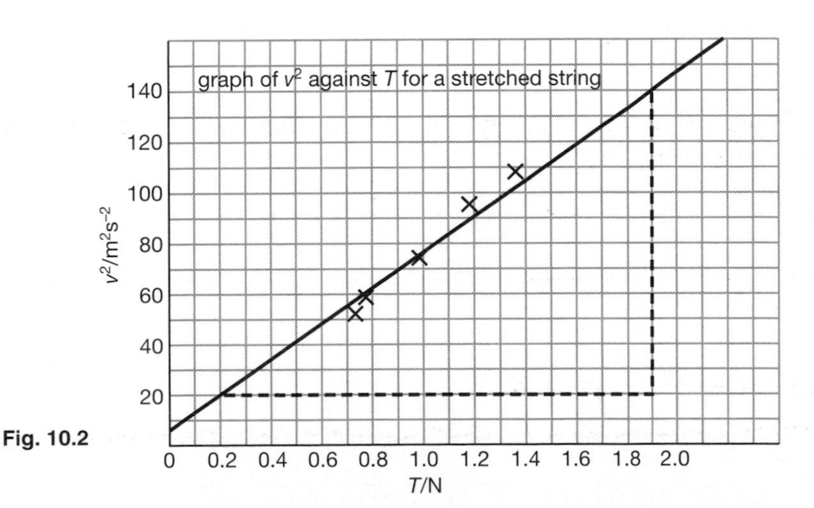

Fig. 10.2

Do the student's results support the relationship given above? Justify your answer.

Yes because they show an almost perfect linear relationship - ✓ ✗

proportional to each other.

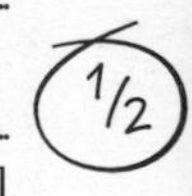

[2 marks]

Use the graph to obtain a value for m.

$$\frac{1}{\mu} = \frac{\Delta(v^2)}{\Delta T} = \frac{120}{1.7} = 70.6 \quad ✓✓✓$$

$$\therefore \mu = 0.014 \text{ kg m}^{-1}$$

[3 marks] 3/3

[Total 9 marks] 4/9

How to score full marks

The candidate answering the question has missed the purpose of the blanks in the table. Instead of calculating values for v^2 and T, she has tried to add data for two more masses with corresponding frequencies and wavelengths.

You need to calculate values of v by multiplying the corresponding values of f and λ, square these values and multiply the corresponding mass values by 9.8 Nkg^{-1} to give values for tension (T). Your completed table would look like this:

M/kg	f/Hz	λ/m	v/ms^{-1}	v^2/m^{2}s^{-2}	T/N
0.16	30.6	0.37	12.3	151	1.96
0.20	30.0	0.41	11.3	128	1.57

The exam candidate has not been able to complete the points on the graph. You would probably have been given a mark for drawing the best straight line through the points already plotted. However no mark has been awarded here because **the line is not the best available since it does not pass through the origin**.

The correct line is shown below.

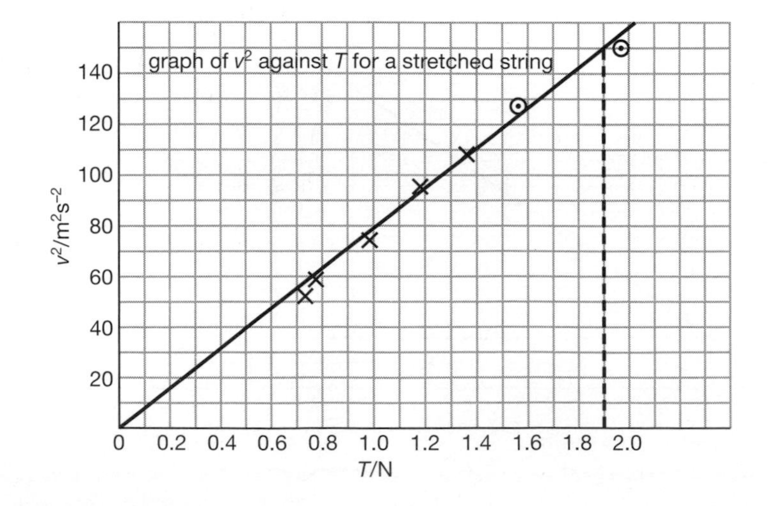

The candidate's statement about how well the results support the relationship is slightly confused, but a single mark is awarded for being partially correct. **The points imply that there is a linear relationship but not *proportionality* since the candidate's line of best fit does not pass through the origin.**

The candidate has used the student's results given in the paper, and her calculation of μ gains all of the last three marks. The candidate has used a large gradient triangle and has correctly manipulated the given equation into the equation for a straight line ($y = mx + c$). The answer clearly demonstrates that the candidate understands the need to do this, but you could show the working:

$$v = \sqrt{\frac{T}{\mu}} \therefore v^2 = \frac{T}{\mu} \therefore v^2 = \frac{1}{\mu}(T)$$

this is equivalent to $y = mx + c$

therefore the gradient = $\frac{1}{\mu}$

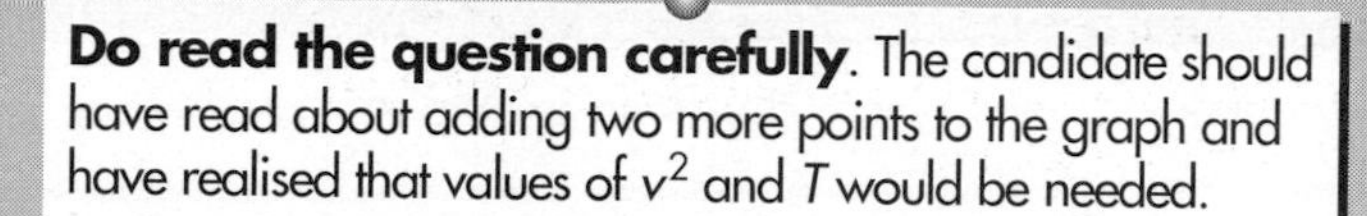

Do read the question carefully. The candidate should have read about adding two more points to the graph and have realised that values of v^2 and T would be needed.

Don't be confused by the terms 'proportional' and 'directly proportional'. These terms mean the same thing! Proportional is an increasingly accepted abbreviation of directly proportional. **To be proportional a straight line must pass through the origin**. If it does not pass through the origin, and is straight, it is simply linear.

Be careful when drawing conclusions relating to data. You aren't expected to comment on what you can't see. In this case the exam candidate's line did not pass through the origin and so it cannot be shown to be proportional – **don't say that the line is proportional if this disagrees with your results.**

Key points to remember

- **Sounds** are **longitudinal waves** consisting of a series of compressions and rarefactions that travel through the medium from the source.
- The audible range consists of frequencies from ≈ 20 Hz to ≈ 20 kHz – frequencies above this are *ultrasonic*.

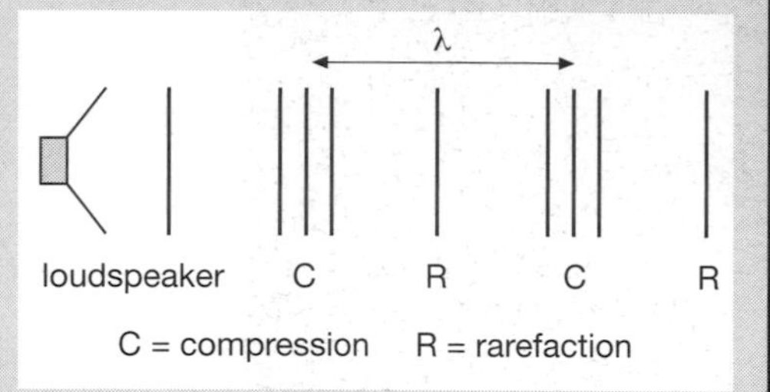

- **Standing waves** occur as a consequence of the superposition of two waves that meet.
- The waves must have equal frequencies and must travel in opposite directions at equal speed.
- The resulting displacement (at any position) is the vector sum of the individual displacements.

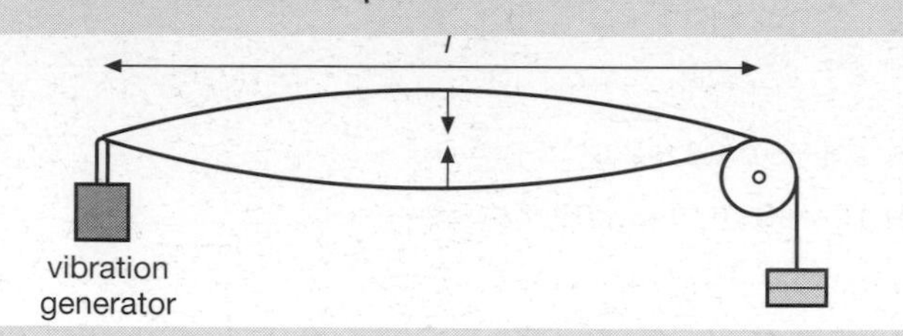

- The frequency of the fundamental standing wave in a string is given by $f = \frac{1}{2l}\sqrt{\frac{T}{\mu}}$

  f = fundamental frequency in Hz, T = tension in N, l = length of string in m, μ = mass per unit length in kgm^{-1}
- The fundamental standing wave in a string can be found only when the frequency of the source matches the natural frequency of the string – this is an example of resonance.

The Doppler effect

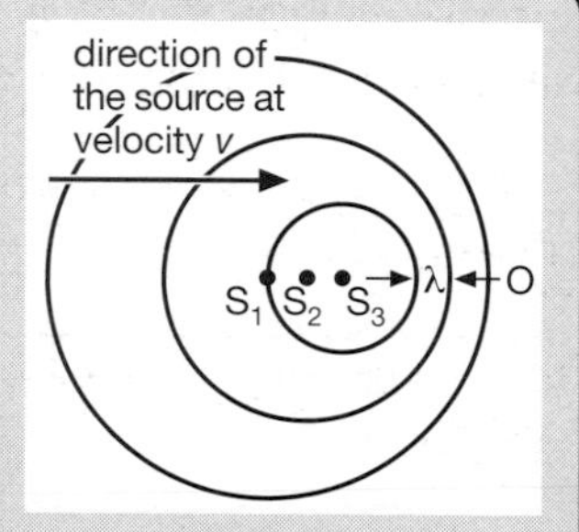

When a sound source approaches an observer, he hears an increase in pitch. When the source moves away from the observer the pitch decreases. As the source moves from S_1 to S_2 to S_3 (towards an observer at O) the effective wavelength decreases. On the opposite side the wavelength increases. This is the Doppler effect.

The Doppler effect applies to light waves as well as sound waves, giving red shifts for light from stars moving away from Earth and blue shifts for those moving towards the Earth.

The change in frequency is given by: $\Delta f = \frac{vf}{c}$

Δf is the change in frequency in Hz, f is the frequency of the source in Hz, v is the velocity of the object in ms^{-1} and c is the velocity of the waves in ms^{-1}. You are unlikely to need to prove the Doppler equation but you should know that:

- the equation only applies when $v \ll c$;
- the effect provides evidence for the theory that the Universe is expanding.

The speed of sound

To measure the speed of sound in free air, a standing wave can be set up by the superposition of incident and reflected waves. The reflector must be adjusted until a maximum is obtained using the oscilloscope/microphone arrangement. The distance between adjacent nodes can be found by measuring the distance the microphone needs to be moved to give successive minima.

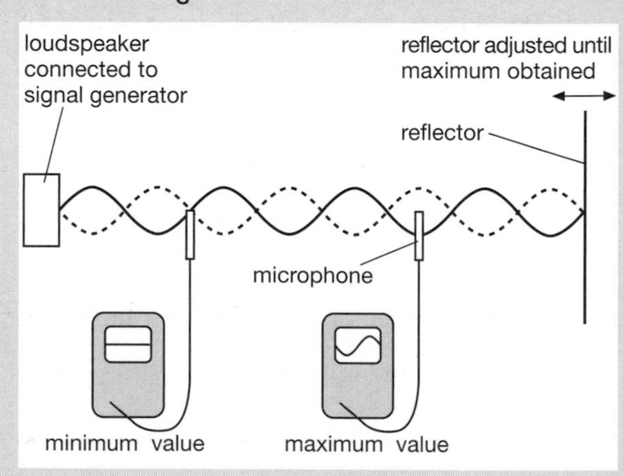

The wavelength is twice the distance between consecutive nodes. The frequency of the sound is read from the signal generator. The speed is then calculated from $c = f\lambda$.

Question to try

Examiner's hints

Part (b)

There are four marks allocated to this and so you should look for four distinct points in the process of setting up this wave.

Part (c)

The total length of the string is given in the stem of the question and so you can calculate the distance between nodes and the wavelength directly from this.

Part (d)

This is a discuss question and so a simple "yes" or "no" is won't gain marks. You need to consider the specific example of the wave on the string produced by the bow and relate your answer to this.

Part (e)

This is not a standard experiment for measuring the speed of sound. You must look at what you can measure from the experiment and what additional information you need to be able to calculate the speed of sound. Once you have done this you can decide which measuring instruments you would need.

There is no unique correct answer to this part, but your method must be workable. Do not invent apparatus or use a vague term like "connect it to a computer"; these things will not earn you credit.

A stretched string on a stringed instrument has a vibrating length of 1.25 m. It is bowed to set it oscillating and it is observed to undergo oscillations, as illustrated in **Fig. 10.4**.

Fig. 10.4

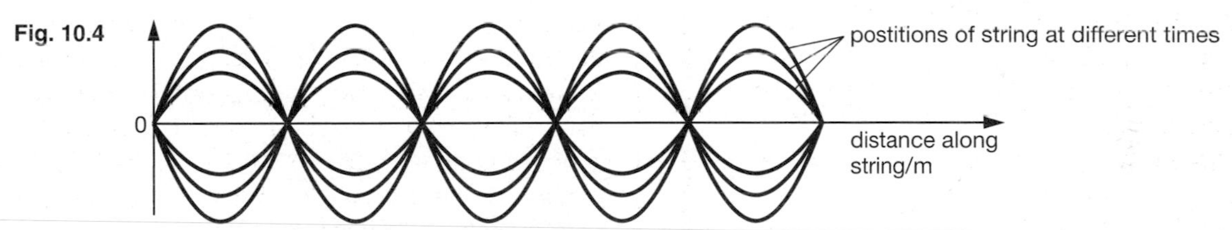

(a) (i) Is this a progressive wave or a standing wave?

....

(ii) Is this wave a transverse wave or a longitudinal wave?

....

[2 marks]

(b) Explain how, as a result of the bow moving the string, this wave is formed.

....

....

....

....

[4 marks]

(c) From **Fig 10.4** deduce

(i) the distance between two nodes,

distance = ... m

(ii) the wavelength of the wave.

wavelength = ... m

[2 marks]

(d) Discuss briefly whether such a wave can be polarised.

....

....

....

....

[3 marks]

(e) Give experimental details of how you would extend the investigation described in this question in order to determine the speed of the wave on the string.

....

....

....

....

....

....

....

....

....

....

....

....

[6 marks]

[Total 17 marks]

The answer to this question is on page 90.

Exam question and student's answer

1 The diagram shows a cross-section through a compact disc.

The metal layer of a CD is the recording surface and contains narrow ridges, which form a spiral around the disc.

Fig. 11.1

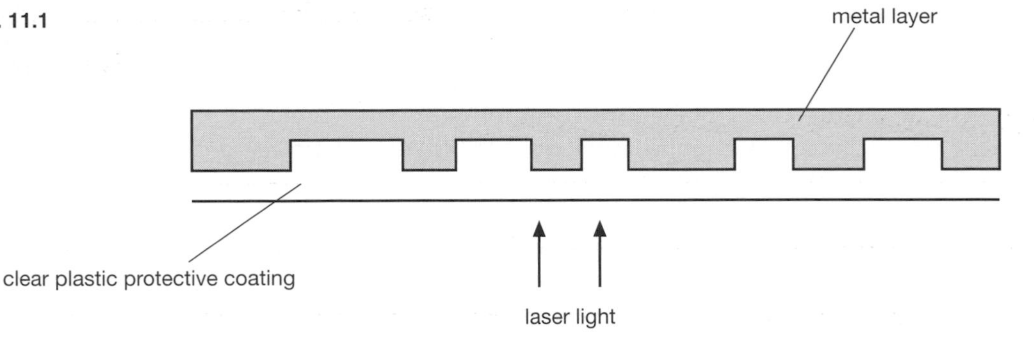

Red *monochromatic* laser light of wavelength 780 nm is used to view these ridges. When the light meets a ridge some of it scatters in all directions and some *interferes destructively* with light reflected from neighbouring valleys.

Explain the meaning of the words in *italics* in the passage above.

Monochromatic of one single colour – only one wavelength ✓ of light – (no different shades of red). (1/1)

Interferes destructively when two or more waves meet, they produce a smaller wave amplitude. (0/1)

(1/2)

[2 marks]

Calculate the frequency of the red laser light.

$c = f\lambda \qquad f = c/\lambda = \dfrac{3 \times 10^8}{780\ \text{nm}}$ ✓

Frequency = 3.85×10^5 Hz ✗ (1/2)

The refractive index of the plastic protective coating is 1.55. What is the speed of the laser light in the plastic coating?

$1.55 = \dfrac{3 \times 10^8}{P} \qquad P = \dfrac{3 \times 10^8}{1.55}$

$= 193548387.1$

Speed = $1.94 \times 10^8\ \text{ms}^{-1}$ ✓ (1/1)

Show that the wavelength of the laser light in the plastic coating is approximately 500 nm.

$$\lambda = c/f = \frac{1.94 \times 10^8}{3.85 \times 10^5} \text{ ecf}$$

1/1

Wavelength = 504 nm ≃ 500 nm ✓

3/4

[4 marks]

The height of the ridges on a CD is approximately 125 nm. Use your last answer to explain how destructive interference occurs.

Since the wavelength of the laser light is longer than the height of the ridges by about four times, it is possible that a wave could reflect off a ridge and meet another wave in such a phase difference as to create destructive interference.

+ =

1/3

[3 marks]

The infrared laser standard was fixed in 1980 because of the reliability and availability of relatively inexpensive lasers, which emit at 780 nm. However, blue light lasers are now being developed. These emit a wavelength of about one half that of the red light lasers.

Will it be possible to play existing CDs using blue light laser CD players? Explain your answer.

No, the height of the ridges would create an incompatability to ✓ the wavelength. The destructive interference would no longer occur in the same way as red light.

[2 marks]

[Total 11 marks]

How to score full marks

The student has gained a mark for a good definition of monochromatic – the bracketed term could easily have been missed out without losing the mark. **Her explanation of destructive interference is not quite full enough.** If she had added a comment about the phase then this could easily have gained full marks. She could have written: "when two waves meet 180° out of phase they tend to cancel each other".

When calculating the frequency of the light, the student seems unsure of the preface to the unit. **"n" stands for "nano"** and = 10^{-9}. This means that the final answer is 3.85×10^{14} Hz.

The refractive index calculation is correct **but it would have been better if the student had written down the equation.** Despite the final answer being both correct and to an appropriate number of significant figures, **there was really no need for her to write down a first version to 10 significant figures!**

The student scores the mark for showing that the wavelength is approximately 500 nm. This is despite not understanding "nm". The error is carried forward, so she is not penalised for the same mistake twice (mark ecf).

The student scores only one out of three for the destructive interference explanation. This is because the **answer does not relate closely enough to the question.** The question says "use your last answer...". The student makes a clear reference to destructive interference, which scores a mark. However, she does not use **the numbers to explain why destructive interference occurs.** Instead of this general answer, you should write: *"When two waves reflect off the top of the ridge and the bottom of the valley the path difference equals twice the height of the ridge (= 250 nm).* ✓ *Since 250 nm =* $\frac{\lambda}{2}$*, the waves are 180° out of phase and interfere destructively."* ✓

The final part of the answer **is too vague to gain both marks.** You need to show the examiner that you recognise that blue light has a wavelength of ≈ 400 nm and so the waves of light falling on the CD and reflecting back from the ridges would now be roughly in phase. This means they would interfere constructively, and so the CD could not be played.

Don't make these mistakes ...

Don't ignore aspects of the question that tell you to do something. "Use your answer to..." means that if you fail to do this you cannot score all the marks. When the answer is numerical then you **must** use those numbers.

Don't write down answers to a large number of significant figures (even as an intermediate step). Such answers are inappropriate because the data used to generate the answer is likely to have been rounded. **Remember the rule of thumb is two or three significant figures.**

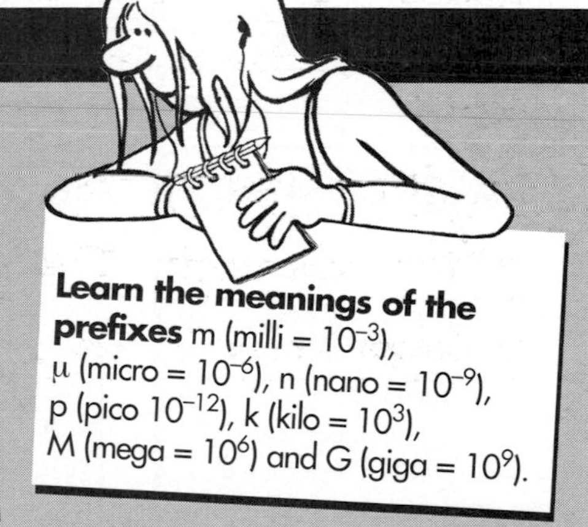

Learn the meanings of the prefixes m (milli = 10^{-3}), μ (micro = 10^{-6}), n (nano = 10^{-9}), p (pico 10^{-12}), k (kilo = 10^{3}), M (mega = 10^{6}) and G (giga = 10^{9}).

Key points to remember

- **Electromagnetic waves** are transverse waves that consist of variations in electric and magnetic fields.
- The speed of light in a vacuum is the maximum speed that anything can travel. $c \approx 3 \times 10^8 \text{ ms}^{-1}$.

The electromagnetic spectrum

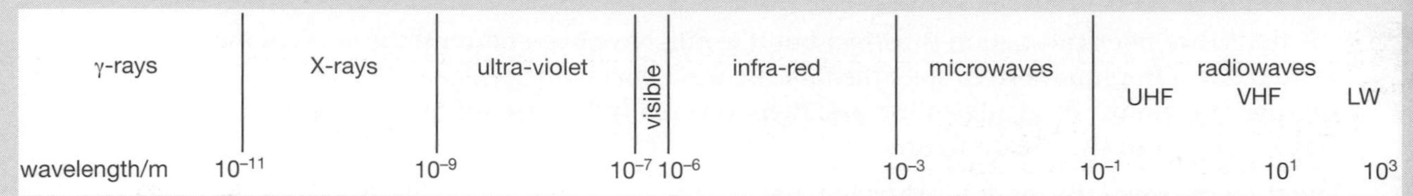

These boundaries are artificial. Waves of identical wavelength may be given two names, depending on the source of the different waves.

- The wavelength of visible light is between ≈ 400 nm and ≈ 700 nm.
- This narrow wavelength range means that a very narrow slit is needed to diffract light noticeably. In practice, less than 0.1 mm will give reasonable diffraction.

  The intensity of the diffracted beam varies:

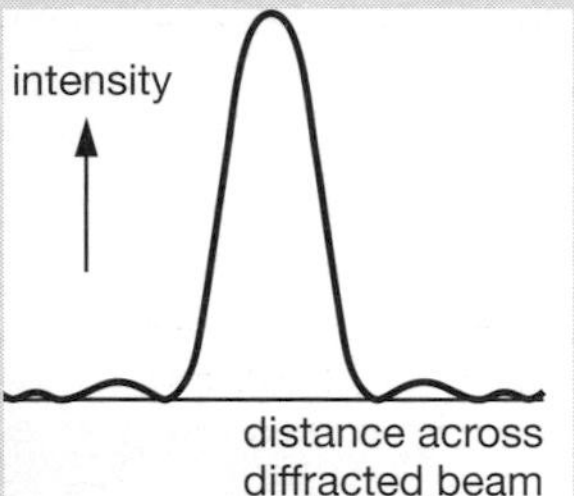

The principal maximum is much more intense and approximately twice as wide as the two secondary maxima.

- **Diffraction gratings** are used to analyse the wavelengths in a particular light.
- Diffracted white light produces several coloured bands.

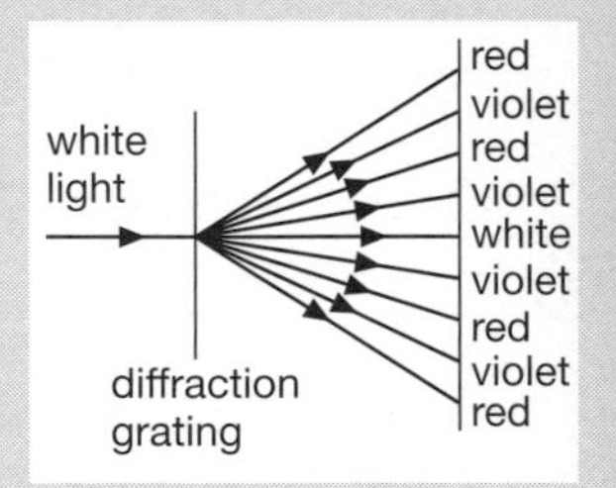

- The diffraction grating equation is $n\lambda = d\sin\theta$ where n is the order of the spectrum (0 – centre, 1 – next on each side etc.), d is the grating spacing (= 1/number of lines per metre) and θ is the angle that a spectrum makes with the straight through position.

- **Young's double slit** for light gives a pattern of equally spaced, alternate bright and dark fringes for waves from a pair of coherent sources – i.e. sources with constant frequency and a constant phase relationship.
- The **fringe separation** is given by $\frac{\lambda D}{d}$ where D is the distance from the slits to the screen and d is the slit separation.

Refraction and Snell's law:

- When light passes from one transparent medium to another it refracts in a predictable way according to Snell's law. where i = angle of incidence, r = angle of refraction and ${}_1n_2$ = a constant called the refractive index when going from medium 1 to medium 2.
- When c_1 = speed of light in medium 1 and c_2 is the speed of light in medium 2 then ${}_1n_2 = \frac{c_1}{c_2}$.

  Snell's Law: $\frac{\sin i}{\sin r} = {}_1n_2$

- In calculations on refraction a useful starting point is that the refractive index, going from air (or vacuum) into a substance, is greater than 1 and light always slows down as it enters a denser medium. So since ${}_1n_2 = \frac{c_1}{c_2}$ when c_1 is the speed of light in a vacuum, medium 2 is the denser medium.
- It is therefore possible to calculate the critical angle for a medium. When light goes from a denser to a less dense medium, it refracts away from the normal ($r>i$). At the critical angle $r = 90°$ so $\sin r = 1$ => $\sin c = {}_2n_1$ (where 2 is the denser medium)

Question to try

Examiner's hints

(a) Five marks tell you that there need to be five distinct points relating to the formation of the bright fringe.

(b) This is really asking you to use the equation for a double slit to relate the fringe separation to three changes.

Q1

Fig 11.2 shows an arrangement to produce interference fringes by Young's two slits method.

Red light of wavelength 7.0×10^{-9} m is being used.

Fig. 11.2

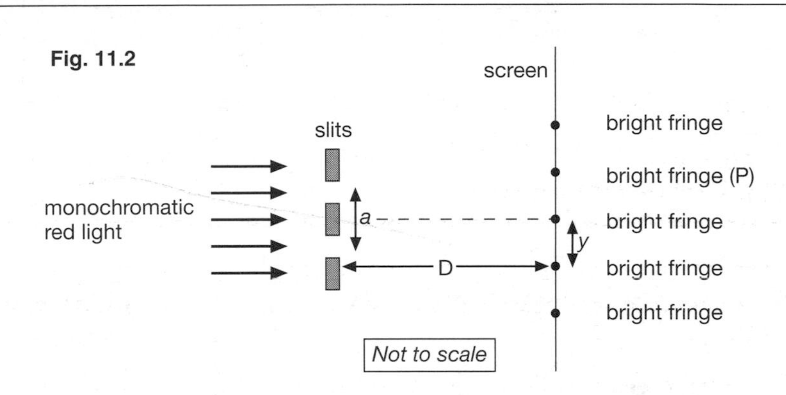

(a) Explain how the bright fringe labelled P is formed.

....

....

....

....

....

[5 marks]

(b) What would be the effect on the fringe width y of separately:

(i) moving the screen further away from the slits;

....

[1 mark]

(ii) increasing the slit separation;

....

[1 mark]

(iii) illuminating the slits with blue light of wavelength 4.6×10^{-9} m?

....

[1 mark]

(c) To obtain an interference pattern, the light from the two slits must be *coherent*. What is meant by the term *coherent*?

....

....

[2 marks]

The answer to this question is on page 90.

[Total 10 marks]

Exam question and student's answer

1 (a) In the photoelectric effect equation

$hf = \phi + E_k$ state what is meant by

hf Total energy 6.63×10^{-34} x frequency given the energy of a wave. ✓

ϕ Work function – energy to release an electron. ✓

E_k Kinetic energy possessed by an electron. ✗ [3 marks]

(b) Monochromatic light of wavelength 3.80×10^{-7} m falls with an intensity of 6.0 μW m^{-2} on to a metallic surface whose work function is 3.2×10^{-19} J.
Using data from the *Data Sheet*, calculate

(i) the energy of a single photon of light of this wavelength,

$$E = hf = \frac{hc}{\lambda} = \frac{6.63 \times 10^{-34} \times 3 \times 10^{8}}{3.8 \times 10^{-7}}$$

$$= 5.2 \times 10^{-19}\ \text{J}\ ✓$$

(ii) the number of electrons emitted per second from 1.0×10^{-6} m^2 of the surface if a photon has a 1 in 1000 chance of ejecting an electron.

power = intensity x area

$= 6 \times 10^{-6}\ \text{W} \times 1.0 \times 10^{-6} = 6 \times 10^{-12}\ \text{W}$

= energy in W (6×10^{-12} ✓ J)

$$\text{number of electrons} = \frac{6 \times 10^{-12}\ \text{J}}{5.2 \times 10^{-19}\ \text{J}}$$

$$= 1.15 \times 10^{7}\ ✗$$

(iii) the maximum kinetic energy which one of these photoelectrons could possess.

$ke = hf - \phi$ ✓

$= 5.2 \times 10^{-19} - 3.2 \times 10^{-19}$

$= 2.0 \times 10^{-19}$ J ✓

4/5

[5 marks]

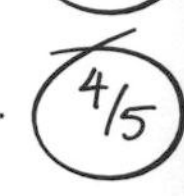

[Total 8 marks]

How to score full marks

art (a):

- Although wave energy has been given the mark for *hf*, the examiner would have expected the student to mention **photon** or quantum of **radiation**. There is no credit for remembering the value of Planck's constant since this is given on the data sheet.
- **Simply mentioning work function would have been enough to gain the second mark** for ϕ. This student has gone on to explain the meaning of work function and therefore has given more detail than required. This is fine providing the extra detail is correct, as it is in this case.
- The student's answer for E_k doesn't score the mark. E_k is the maximum kinetic energy of the electrons emitted when a photon penetrates deep into the metal before interacting with an electron in the metal; some electrons have less energy than this. This student has also been imprecise by talking about "an electron". **The equation refers very specifically to a photoelectron released by the photoelectric effect**.

arts (b) (i) and (iii):

hese parts are well answered and very clearly set out.

art (b)(ii):

his answer is nearly completely correct but there are only two marks for this part of the question, o a less than perfect answer only gains one mark. This student lost the other mark for not emembering that **only 1 in 1000 photons ejects an electron**. The student's answer needs to be ivided by 1000 to give the correct answer of 1.15×10^4.

ntensity has units of Wm^{-2}

Don't make these mistakes ...

Don't give more detail than you need – in case you are wrong! You should **recognise that you do not need to explain your answer when you see the keyword "state"**.

Avoid missing out words such as maximum or minimum when you revise your equations. If you learn Einstein's photoelectric equation as:

$$E = \Phi + (\tfrac{1}{2}\, mv^2)_{max}$$

you are unlikely to forget that this is the maximum kinetic energy of the emitted electrons.

Try to develop the habit of reading questions thoroughly and then **asking yourself about the question set**. For example, in this question you might ask, "What is the significance in the examiners telling me that only 1 in 1000 photons release electrons?" **This should remind you to divide by 1000 in the end**.

Key points to remember

- **Electrons** in atoms occupy shells.
- Each shell has an associated energy value.
- When an electron moves from one energy level to another it must gain or give out energy equal to the difference between the energy levels.
- Energy absorbed or emitted is quantized in values = hf (where h is Planck's constant and f is the frequency of the radiation)

The **electron volt (eV)** is a unit of energy used in atomic and nuclear physics. 1eV = 1.6×10^{-19} J. It is the kinetic energy gained by an electron as it is accelerated through a p.d. of 1 V.

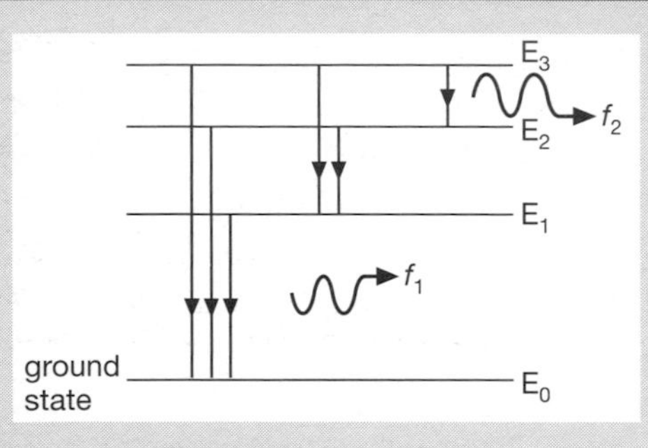

- The **ground state** is the most stable state of the atom – the lowest energy level. This is the state in which the electron has least energy.
- The greater the difference between the energy levels the greater the energy (and the higher the frequency) of the emitted photon.

$$E_1 - E_0 = hf_1 \text{ and } E_3 - E_2 = hf_2$$

$$\therefore \; f_2 < f_1$$

- When an electron, in the ground state, absorbs exactly the right amount of energy it can jump to a higher energy level. The electron has been excited.
- Electrons falling from higher to lower energy levels produce an emission spectrum.
- Analysis of **emission spectra** gives information about the nature of the element undergoing transition.
- Hot solids, liquids and high density gases (such as stars), emit a **continuous spectrum**.

Red Orange Yellow Green Blue Indigo Violet

- A **line emission spectrum** is produced when gases made up of individual atoms at low pressure de-excite and lose energy.

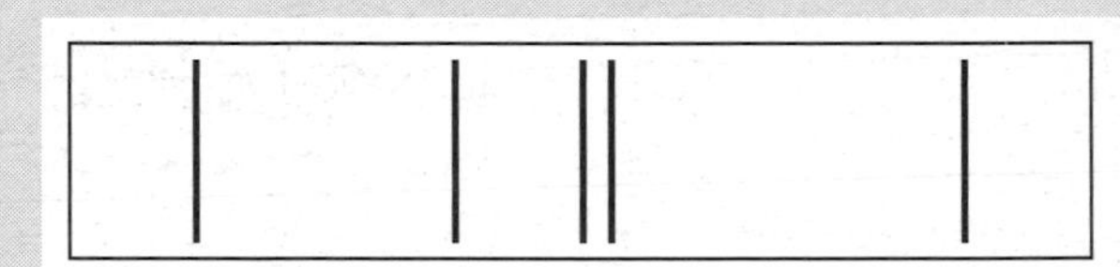

- A **line absorption spectrum** consists of dark lines in the same position as those in the emission spectrum but against a continuous spectrum.
- A **band spectrum** is emitted by gases in which the atoms are close enough to influence each other (O_2, H_2 etc.).

Wave-Particle Duality

The **photoelectric effect** is largely responsible for the view that radiation must be considered as both a wave and a particle.

Shining more intense light on a metal surface does not produce any photoelectrons unless the frequency of the light is high enough to overcome the work function of the metal. If the frequency is high enough, then increasing the intensity does not release more energetic electrons as we would expect with a wave. Instead more electrons are released, each with kinetic energy up to the allowed maximum.

De Broglie applied the wave theory to "particles" attributing a wavelength λ to them such that $\lambda = h/p$ where h is the Planck constant and p is the momentum of the particle. Electron diffraction proves that "particles" can exhibit wave properties.

Photoelectric effect

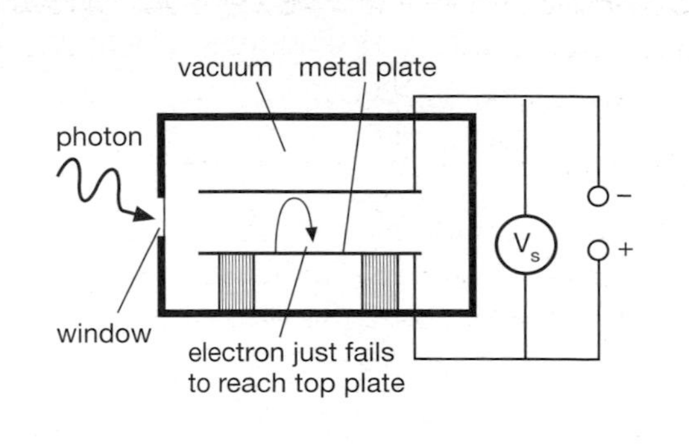

- When light shines on to some metals, electrons may be emitted according to Einstein's photoelectric equation:

  $$E = \Phi + (\tfrac{1}{2}\ mv^2)_{max}$$

  Where E = photon energy (=hf), Φ is the work function energy of the metal and $(\tfrac{1}{2}\ mv^2)_{max}$ is the maximum kinetic energy of the emitted electrons.

  Φ is the minimum energy required for an electron to escape from the surface of the metal (=hf_0 – where f_0 is the threshold frequency – the minimum frequency which will allow electron emission).

- The maximum kinetic energy of the emitted electrons can be calculated from the "stopping voltage" (V_s) of the evacuated photocell.

  $$eV_s = (\tfrac{1}{2}\ mv^2)_{max}$$

Question to try

Examiner's hints

Parts (a) and (b): Try to make a single point for each of the marks that are available. In part (b) it essential that you relate the emission to the absorption spectrum so you will need to mention both of them.

Part (c)(i): Again you will need to mention both emission and absorption spectra in your answer but start each time with a separate sentence.

Part (c) (ii): If you feel that a sketch would help your answer here, use a space somewhere on the page and draw the examiner's attention to the fact that you are doing this.

Part (d) (i): You need to relate the wavelength to the difference in energy between the energy levels. Convert electron volts to joules using the fact that the charge on an electron is 1.6×10^{-19} C. Planck's constant = 6.6×10^{-34} Js.

Part (d) (ii): The question here implies that you do not need to explain your answer.

Part (e) (i): This part can either be done by working out the wavelength of the radiation and relating it to the electromagnetic spectrum or else you may simply know what sort of radiation a mercury vapour lamp emits.

Part (e) (ii): The ionisation energy is the energy needed to remove an outer electron from the atom. It is equivalent to the energy of the highest frequency photon. Don't forget to convert your answer from joules into eV.

Q1

(a) Describe the physical appearance of

(i) *a line emission spectrum*

....

.... [2 marks]

(ii) *a line absorption spectrum*

....

.... [2 marks]

(b) Describe the relationship between the two spectra named in **(a)**

.... [1 mark]

(c) **(i)** Explain, at the atomic level, how *line emission* and *line absorption spectra* originate.

....

....

.... [4 marks]

(ii) Describe **briefly** how you would demonstrate a line absorption spectrum.

....

.... [3 marks]

(d) The diagram below shows some of the energy levels of an atom,

–1.89 eV

–2.06 eV

–2.37 eV

–3.03 eV

(i) Calculate the longest wavelength in the line spectrum arising from these levels.

....

....

....

[4 marks]

(ii) What is the maximum number of lines possible from these four energy levels?

....

[1 mark]

(e) The highest frequency light emitted from a mercury discharge lamp is 2.5×10^{15} Hz.

(i) In which part of the spectrum is this frequency?

....

[1 mark]

(ii) Calculate the ionisation energy of a mercury atom in eV.

....

[2 marks]

The answer to this question is on page 91. [Total 20 marks]

Exam question and student's answer

1 Name *two* sources of natural background radiation.

1 Rocks, eg granite ✓

2 Universe/space ^

[2 marks]

Caesium-137 is a by-product of nuclear fission within a nuclear reactor. The nuclear equation below describes the production of $^{137}_{55}Cs$. Complete the two boxes in **Fig 13.1**:

Fig. 13.1

$$^{235}_{92}U + ^{1}_{0}n \rightarrow ^{137}_{55}Cs + \boxed{^{95}_{37}Rb} + \boxed{3^{1}_{0}n}$$

✓ ✗

[2 marks]

The half-life of caesium-137 is 30 years. When the fuel rods are removed from a nuclear reactor core, the total activity of the caesium-137 is 6.4×10^{15} Bq.

Explain the phrase *the activity of the caesium-137 is* 6.4×10^{15} *Bq*.

6.4×10^{15} caesium-137 nuclei decrease ✗ per second. ✓

[2 marks]

After how many half-lives will this activity have fallen to 2.5×10^{13} Bq? Explain your working.

$6.4 \times 10^{15} = 2.5 \times 10^{13} \times 2^x$

$\therefore 2^x = 256 \; (= 2 \times 2 \times 2 \times 2 \times 2 \times 2 \times 2 \times 2)$ ✓

$\therefore x = 8$ ✓

Number of half-lives = 8

[2 marks]

Comment on the problems of storage of the fuel rods over this time period and beyond.

This is a long time and they must ✓ be safely contained in this time. ✗

Keeping the rods away from people over this time is expensive. ^

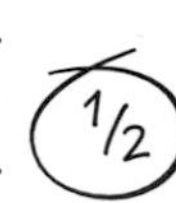

[2 marks]

[Total 10 marks]

How to score full marks

- The first part of the answer is quite reasonable but **the student was a little vague when talking about "Universe/space"**. You could say, "cosmic rays, which come from space". An even better answer would be "radon gas", which is formed in the ground and is responsible for up to 50% of the background radiation.
- The nuclear equation is nearly completely correct but the student has forgotten the incident neutron (on the left-hand side of the equation) and so **there should be four not three neutrons on the right hand side of the equation**.
- Again the explanation of activity is essentially correct but leaves sufficient doubt in the examiner's mind for the answer not to gain full marks. **On decaying the nuclei do not disappear they simply become new nuclei**. In a sense, the number of **caesium** nuclei do decrease because they have become nuclei of a new element. However the number of nuclei is still the same. A better and unambiguous answer is "6.4×10^{15} nuclei per second decay into a new element".
- The next part of the answer gains full marks despite the student using **no words to explain the calculation**. This is set out so **clearly** that it is very easy to follow mathematically and this is just as good as using words to explain the answer.
- The last part of the answer is certainly along the right lines but is again rather vague and cannot score full marks. **The student has not clearly explained his/her reasoning**. *What is a long time? What must be safely contained? Why must the fuel rods be kept safely? "Safe" from what or for whom?*
- **A good answer would be any of the following**:
  - A suitable storage facility would need to be sited free from exposure to weather (or landslip) to prevent the fuel rods becoming exposed by erosion and putting the public at risk.
  - The containers would need to be corrosion-safe and well away from watercourses so that there would be no risk of contamination of drinking water.
  - People charged with looking after the fuel rods would need to be able to monitor them over a long period of time because of their relatively long half-life.
  - The fuel rods are going to be highly radioactive for a very long period of time and so it is essential that they are stored well away from humans and animals so that neither can have access to them either by mistake or design.

Don't make these mistakes ...

In a question asking you to discuss or comment on something do **make sure that your answer is actually linked to the question**. Vague answers rarely gain full marks.

Don't forget that the quality of your written communication will be assessed by the answers you give. **Each of the specifications tell you how this assessment is done and you should make sure you read these.** On your more descriptive answers try to ensure that your spelling, punctuation and grammar are of a high standard.

Key points to remember

- **Radioactivity** is a random process in which nuclei emit ionizing radiation. It occurs spontaneously in certain atomic nuclei.
- The nucleus consists of positively charged protons and uncharged neutrons.
- The notation ${}^{A}_{Z}X$ is used to describe the nucleus of element X. Z is the proton number and A the nucleon number ∴ A–Z is the number of neutrons.

- **Alpha emission** is described by:

$$ {}^{A}_{Z}X \longrightarrow {}^{A-4}_{Z-2}Y + {}^{4}_{2}He \text{ (α-particle)} $$

- **Beta emission** is described by:

$$ {}^{A}_{Z}X \longrightarrow {}^{A}_{Z+1}Y + {}^{0}_{-1}e \text{ (β-particle)} $$

- **Gamma emission** has no effect on the structure of the nucleus. A gamma rays is emitted when the original nucleus is left in an excited state after the emission of an α or β particle:

$$ {}^{A}_{Z}X^{*} \rightarrow {}^{A}_{Z}X + {}^{0}_{0}\gamma \text{ (photon)} $$

*denotes the excited nucleus

Radioactive decay

- Radioactive decay is exponential and so never falls to zero. We cannot, therefore, talk about the lifetime of a sample of material.
- The half-life ($T_{1/2}$) is the time taken for the activity of a sample of radioactive material to fall to half its original value.
- A graph of activity against time is the same shape as a graph of the number of unstable nuclei present against time.

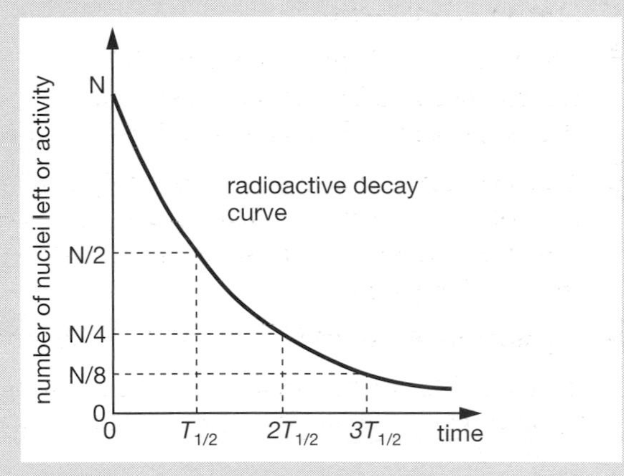

- The activity is related to the number of nuclei remaining by $A = -\lambda N$.
- λ is the decay constant and is $= 0.69/T_{1/2}$ it has units of time^{-1} (e.g. s^{-1}).

alpha particle	beta particle	gamma rays
2 protons + 2 neutrons	fast moving electron	very short wavelength electromagnetic waves
range of a few cm in air	range of several decimetres in air	infinite range in air – obeys the inverse-square law
very strong ionizer	quite strong ionizer	very weak ionizer
absorbed by thin paper or skin	absorbed by a few mm of aluminium or several cm of flesh	not completely absorbed even by several metres of concrete or several cm of lead
charge +2e	charge –1e	uncharged

Key points to remember

Alpha scattering

You should know how important this experiment was in determining the structure of the atom:

- Rutherford, Geiger and Marsden were able to demonstrate that the atom consists of a nucleus surrounded by electrons. They did this by studying the scattering alpha particles aimed at a very thin gold foil.
- Most alpha particles pass straight through the foil showing that the majority of the atom is empty space.
- A small number of alpha particles slightly deviate from their forward motion but significantly about 1 in 8000 is "scattered" through an angle of more than 90°, ie. back-scattered.
- These experiments showed that the nucleus carries a positive charge and it has a diameter of $\sim 10^{-15}$ m compared with the atom, which has a diameter of $\sim 10^{-10}$ m. Thus, if the nucleus were a marble of diameter 1 cm, the electrons would occupy a sphere of diameter 1 km on the same scale!

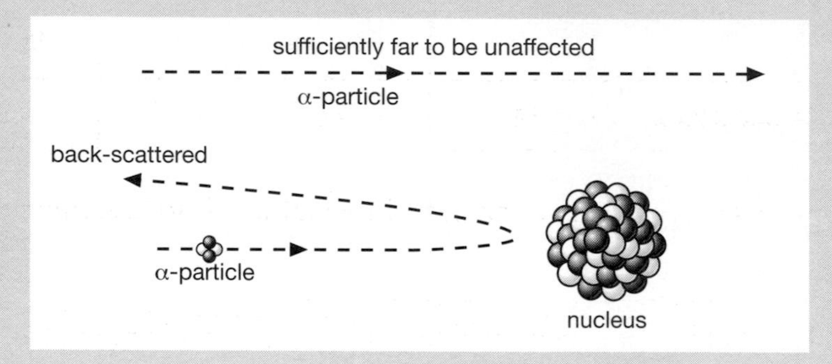

Questions to try

Examiner's hint for question 1
- This question gives the mean background count rate and so you need to correct the readings before plotting them.
- It's usually a good idea to calculate the half-life at least twice and find the mean. This will tell you if you are doing something wrong (because your answers may be very different).

Q1

The table shows the count rate from a radioactive isotope. The background count rate is 0.4 counts s^{-1}.

Time, *t* / *s*	0	30	60	90	120
Count rate / counts per second	14.0	9.6	6.6	4.6	3.2

Plot a graph of the **corrected count rate** against time and use it to determine the half-life of the isotope.

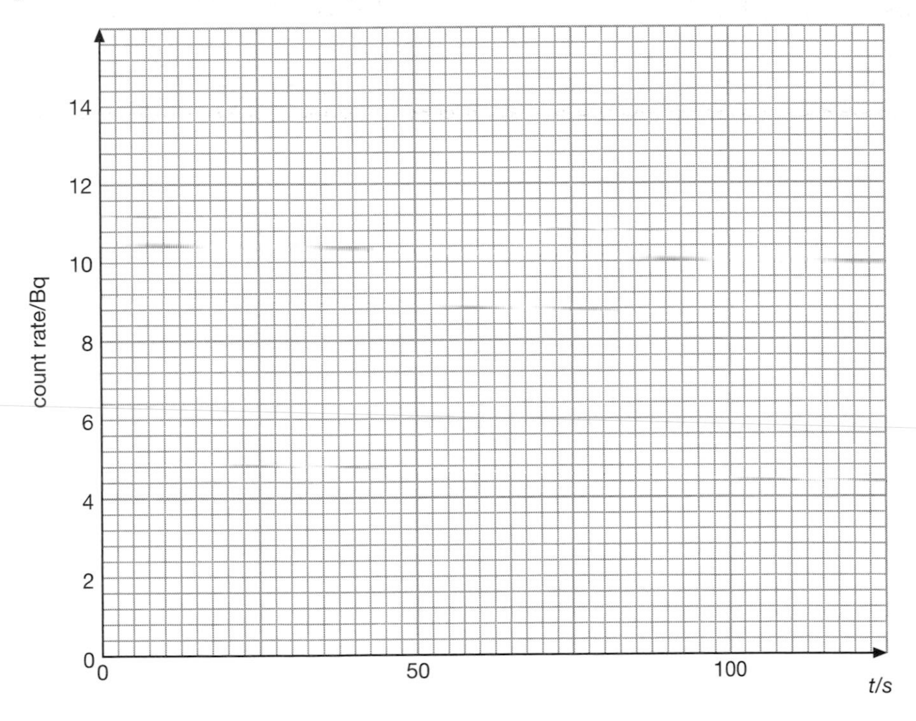

Half-life ....

[Total 5 marks]

Examiner's hints for question 2

Part (a): Five marks are awarded here so the examiners are looking for five clear points – don't worry if you can think of more than five because it is likely that a variety of answers will be acceptable.

Part (b): By "large angles", the question is really talking about angles greater than 90°. Don't forget to mention both the structure of the atom and the forces acting on the α particle.

Q2

(a) Describe the principal features of the *nuclear model* of the atom suggested by Rutherford.

....

....

....

....

....

[5 marks]

(b) When gold foil is bombarded by α particles it is found that most of the particles pass through the foil without significant change of direction or loss of energy. A few particles are deviated from their original direction by more than 90°. Explain, in terms of the nuclear model of the atom and by considering the nature of the forces acting,

(i) why some α particles are deflected through large angles,

....

....

....

....

(ii) why most of the α particles pass through the foil without any significant change in direction.

....

....

....

....

[5 marks]

[Total 10 marks]

The answers to these questions are on pages 91 and 92.

Exam question and student's answer

1 A particle accelerator is designed to accelerate antiprotons through a potential difference of 2.0 GV and make them collide with protons of equal energy moving in the opposite direction. In such a collision, a proton-antiproton pair is created, as represented by the equation:

$$p + \bar{p} \rightarrow p + \bar{p} + p + \bar{p}$$

You may assume that the rest energy of the proton is 940 MeV.

(a) State how an antiproton differs from a proton.

An antiproton has opposite charge and is made up of $\bar{u}\bar{u}\bar{d}$ rather than uud. ✓

[1 mark]

(b) Give the total kinetic energy of the particles, in GeV, before collision.

4.0 GeV ✓

[1 mark]

(c) State the rest energy of the antiproton.

940 MeV ✓

[1 mark]

(d) Calculate the total kinetic energy of the particles, in GeV, after the collision.

4.0 GeV − 940 MeV

= 3.1 GeV ✓ ✗ ✓

[3 marks]

[Total 6 marks]

How to score full marks

Part (a)

A more common answer would be, "although each particle has the same rest mass, the proton carries a positive charge while the antiproton carries an equal negative charge".

The student's answer – that the particles have opposite charges – was not good enough. **The answer is more than salvaged by giving the quark constituents of protons and antiproton** (proton – two up quarks and one down quark; antiproton – two anti-up quarks and one anti-down quark).

Part (b)

The student has realised that **the total kinetic energy is the sum of the kinetic energies of the proton and the antiproton**. This is one place where there would not be a unit penalty since the question tells you what units to use. However the student appears to have recognised that a proton (of charge +e) being accelerated through a p.d. of 2 GV gains kinetic energy 2 **GeV**.

Part (c)

The rest energy of the antiproton is the same as that of the proton, since they also have the same rest masses.

Part (d)

This answer is not full enough and so we cannot see whether the mistake is a careless slip or lack of understanding. Since a second proton-antiproton pair is being created from the kinetic energy of the colliding pair, 2 × 940 MeV is needed to create the particles (i.e. their rest masses). This leaves (4.0 – 1.9) GeV = **2.1 GeV** – the correct answer. This extra energy will be shared between the four particles resulting from the interaction. The student has gained two marks because he/she has subtracted the rest mass of a **single proton or antiproton** and has correctly converted 940 MeV into GeV.

Don't make these mistakes ...

Don't worry if your answers don't always correspond to the mark schemes that are published by your awarding body. The published mark scheme has to be brief enough to be useable for centres but examiners do consider many extra variants of answers at a co-ordination meeting after the examination. Candidates often come up with answers that are correct physics, which surprises the examiners because they had not considered the candidate's approach when compiling the question. **If your physics is correct and your answer is complete, you will be given suitable credit, even if it is not the 'expected' answer**.

Do make sure that you use the units required by the question. You may have thought that the kinetic energy should have been in joules but in atomic physics energies are usually measured in multiples of electron volts (eV). Rest masses are often measured in GeV/c^2, which is convenient, if a little quirky!

Key points to remember

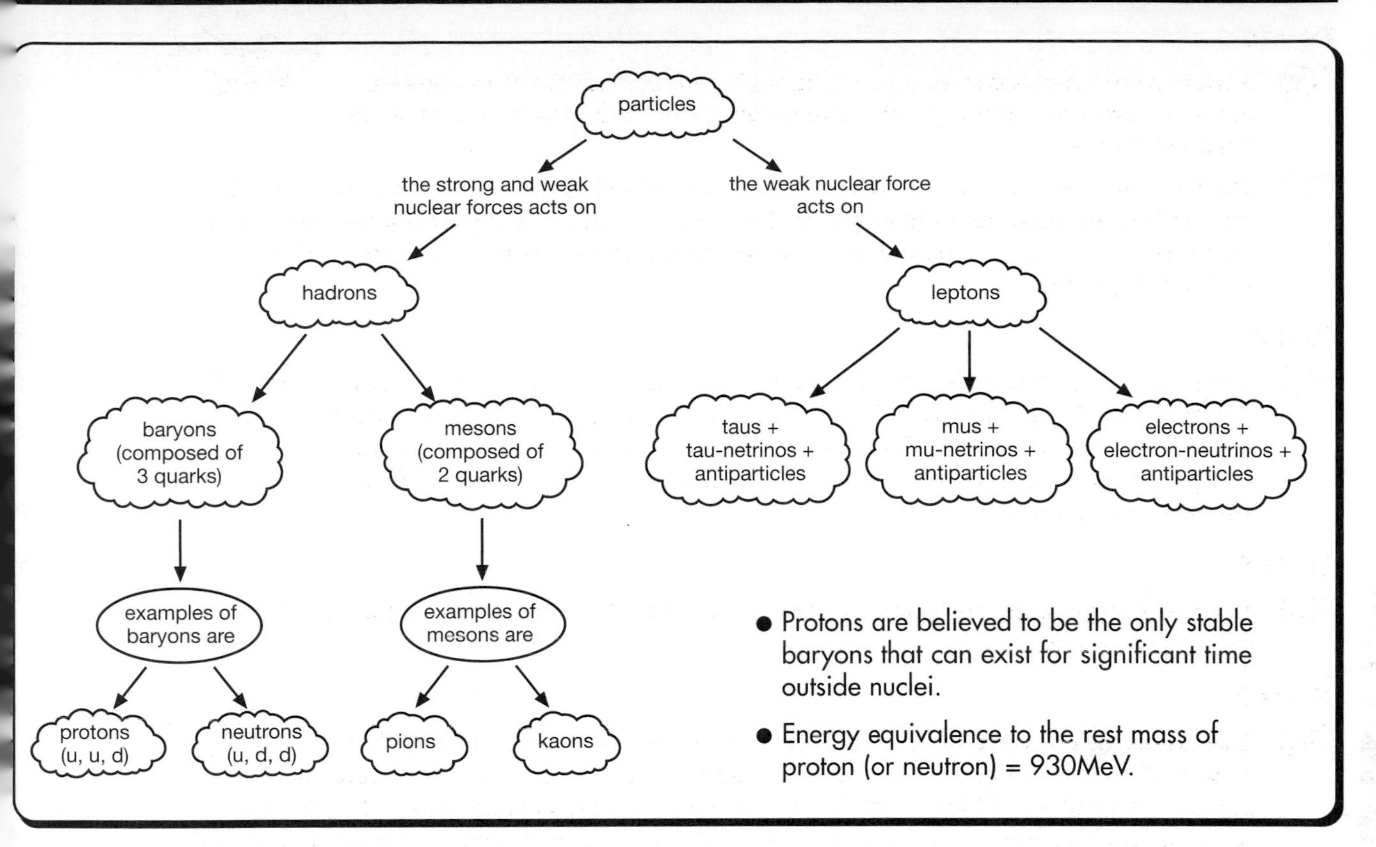

- Protons are believed to be the only stable baryons that can exist for significant time outside nuclei.
- Energy equivalence to the rest mass of proton (or neutron) = 930MeV.

- Each **particle** has an **antiparticle** – having some properties (like mass and spin) that are identical to the particle and some properties (such as charge) that are of the opposite sign.
- The **positron** (e^+) is an **anti-electron**. The positron has the same mass and size of charge as an electron but the positron carries a positive charge whilst the electron carries a negative charge.

Conservation laws:

- Any possible interaction must obey the conservation laws of charge, baryon number, strangeness, and lepton number.
- The laws will not prove that an interaction takes place – they can only confirm it is a possibility.

Fundamental interactions

Interaction	Gravitation	Strong nuclear	Weak nuclear	Electro-magnetic
Particles acted on	all types	quarks gluons*	quarks leptons	charged particles
Property acted on	mass-energy	colour charge	flavour	electrical charge
Exchange particle	graviton (not yet discovered)	gluon	W^+, W^- Z^0 bosons	γ (photon)
Range/m	infinite	$\approx 10^{-15}$	$\approx 10^{-18}$	infinite

*the strong interaction can also be considered to act on hadrons

Key points to remember

Some quark properties

quark	charge/e	Baryon number
u	$+\frac{2}{3}$	$+\frac{1}{3}$
$\bar{u}$	$-\frac{2}{3}$	$-\frac{1}{3}$
d	$-\frac{1}{3}$	$+\frac{1}{3}$
$\bar{d}$	$+\frac{1}{3}$	$-\frac{1}{3}$
s	$-\frac{1}{3}$	$+\frac{1}{3}$
$\bar{s}$	$+\frac{1}{3}$	$-\frac{1}{3}$

Some lepton properties

particles	lepton numbers		
	electron	mu	tau
e^- and ν_e	+1		
e^+ and $\bar{\nu}_e$	−1		
μ^- and ν_μ		+1	
μ^+ and $\bar{\nu}_\mu$		−1	
τ^- and ν_τ			+1
τ^+ and $\bar{\nu}_\tau$			−1

- The six **leptons** (and their six antiparticles) are considered fundamental, as are the 6 **quarks** (up, down, top, bottom, strange, charm) and their antiparticles.

β- decay

- Beta-particles are emitted resulting from:
  neutron → proton + electron + antineutrino.
- Alternatively we can view this as one of the down quarks in the neutron changing into an up quark and emitting the electron (as beta particle) and the anti-neutrino.
- You should make sure that you can see why charge, baryon number and lepton number are each conserved.

Questions to try

Examiner's hints for question 1

Part (a)(i) and (ii): Part (i) just requires a statement; part (ii) is the explanation. Don't give more detail than you are required to – it is a waste of your valuable time!

Part (b)(i): Some specifications would ask for this to be shown on a Feynman diagram. You are not expected to do that here but a correct Feynman diagram would probably be accepted.

Part (b)(ii): Don't forget to explain your answer here.

Q1

(a) Up quarks have a charge of $+\frac{2}{3}e$ and down quarks have a charge of $-\frac{1}{3}e$.

(i) State the number of each type of quark in a neutron.

....

[2 marks]

(ii) Explain in terms of charge why a neutron has this composition.

....

[1 mark]

(b) (i) A neutron decays by β emission. Complete the following decay equation naming all the particles produced in the decay.

neutron → β (electron) + ...

[2 marks]

(ii) State and explain the change of quarks which occurs when this decay happens.

....

....

[2 marks]

[Total 7 marks]

Q2

(a) State which interaction, strong or weak, is experienced by each of the following particles.

hadrons: ....

leptons: ....

[2 marks]

(b) Give **one** example of a hadron and **one** example of a lepton.

hadrons: ....

leptons: ....

[2 marks]

(c) Hadrons are classified as either baryons or mesons. How many quarks are there in a baryon and in a meson?

baryon: ....

meson: ....

[2 marks]

(d) **(i)** State the quark composition of a neutron.

....

(ii) Describe, in terms of quarks, the process of β^- decay when a neutron changes into a proton.

....

....

....

(iii) Sketch a Feynman diagram to represent β^- decay.

[4 marks]

The answers to these questions are on page 92.

[Total 10 marks]

Exam question and student's answer

1 The passage below is taken from the marketing material supplied by a manufacturer of electrically heated showers.

Most electric showers draw cold water direct from the main supply and heat it as it is used – day or night. Not only are they particularly useful for those who do not have a stored hot water supply, they are versatile because every home can have one.

Write a word equation to describe the energy/transfer that takes place in an electric shower.

The energy from the current passing through the wires is changed to internal energy in a heater. The heat then passes to the water, raising ✓ its temperature. This then warms the person having the shower.

1/2

[2 marks]

Rewrite the equation using the appropriate formulae.

$ItV = ms\Delta T$ ✓

1/1

[1 mark]

The technical data supplied by one manufacturer states that their most powerful shower system is fitted with a 10.8 kW heating element and can deliver up to 16 litres of water per minute.

Show that the showering temperature is about 25 °C if the temperature of the mains water is 15 °C and the shower is used at its maximum settings.

(Specific heat capacity of water = 4200 J kg^{-1} K^{-1}
Mass of 1 litre of water = 1.0 kg.)

$\frac{10\,kg}{60\,s} = 0.267\,kg\,s^{-1}$

$10800\,J\,s^{-1} \times 1\,s = 10800\,J$

$\frac{10800\,J}{0.267\,kg} = 40500\,J\,kg^{-1}$

$\frac{40500\,J\,kg^{-1}}{4200\,J\,kg^{-1}\,K^{-1}} = 9.6\,K \simeq 10\,K$

$15\,°C + 10\,K = 25\,°C.$ ✓✓✓

3/3

[3 marks]

The marketing material includes the statement:

Please remember that during the colder months, flow rates may need to be reduced to allow for the cooler temperature of incoming cold water.

Calculate the flow rate required for an output of 40 °C when the incoming water temperature is 10 °C.

$40 - 10 = 30\,°C$

$30\,K \times 4200\,J\,kg^{-1}\,K^{-1} = 126\,000\,J\,kg^{-1}$

$\therefore \frac{126000\,J}{10800\,J\,s^{-1}} = 11.7\,s$

$\frac{1\,kg}{11.7\,s} = 0.086\,kg\,s^{-1}$

$= 5.14\,kg\,min^{-1}$ ✓✓

Flowrate = $5.14\,kg\,min^{-1}$ 2/2

[2 marks]

The maximum steady current drawn by the unit is about 45 A. However, when the shower is first turned on the current is much higher for a short time. Suggest a possible explanation.

The shower has an operating temperature that is much higher than the starting temperature. A 45 A current quickly brings the temperature to its working level, then it cuts back to prevent burn-out. ✗

[1 mark]

[Total 9 marks] 7/9

How to score full marks

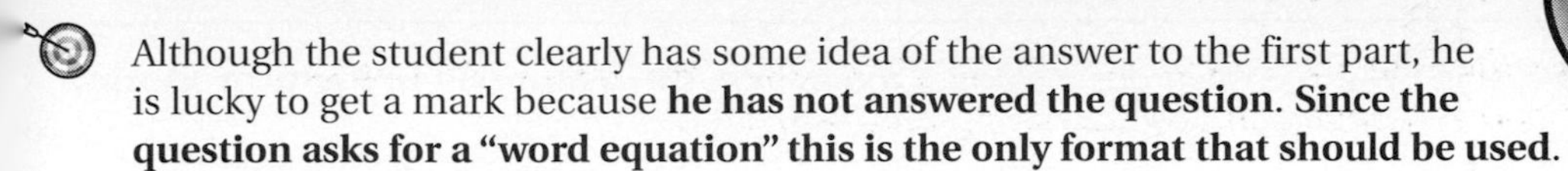

Although the student clearly has some idea of the answer to the first part, he is lucky to get a mark because **he has not answered the question. Since the question asks for a "word equation" this is the only format that should be used.**

The expected answer is:
electrical energy → heat (or better *internal energy*) of the water.

The student **uses the correct formulae** to score the mark in the second part.

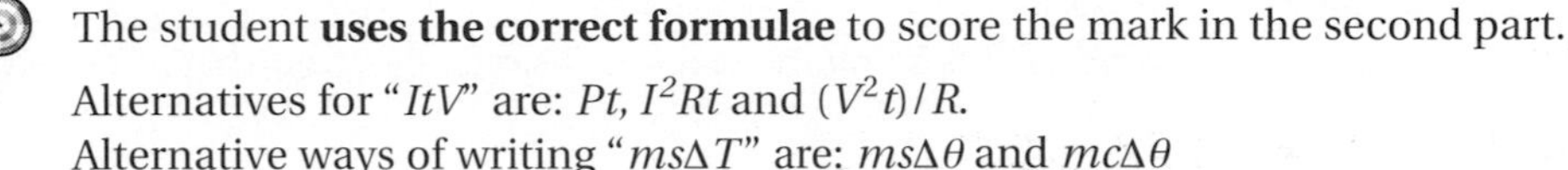

Alternatives for "ItV" are: Pt, I^2Rt and $(V^2t)/R$.
Alternative ways of writing "$ms\Delta T$" are: $ms\Delta\theta$ and $mc\Delta\theta$

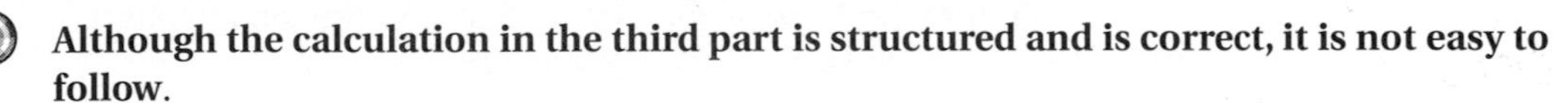

Although the calculation in the third part is structured and is correct, it is not easy to follow.

An answer that explains the working a little more fully is:

$Pt = mc\Delta\theta$

$10.8 \times 10^3\,\text{W} \times 60\,\text{s} = 16\,\text{kg} \times 4200\,\text{J kg}^{-1}\,\text{K}^{-1} \times \Delta\theta$

$$\Delta\theta = \frac{6.48 \times 10^5\,\text{J}}{6.72 \times 10^4\,\text{J K}^{-1}} = 9.6\,\text{K}$$

With the initial temperature being 15 °C

Final temperature = 24.6 °C ≅ 25 °C

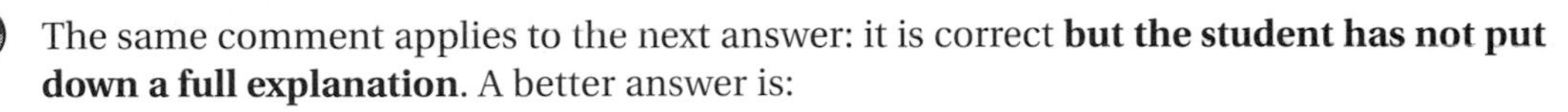

The same comment applies to the next answer: it is correct **but the student has not put down a full explanation**. A better answer is:

$Pt = mc\Delta\theta$

$$\therefore \frac{m}{t} = \frac{P}{c\Delta\theta}$$

$$= \frac{10.8 \times 10^3\,\text{W}}{4.2 \times 10^3\,\text{Jkg}^{-1}\,\text{K}^{-1} \times 30\,\text{K}}$$

$= 8.6 \times 10^{-2}\,\text{kg s}^{-1}$

$= 5.14\,\text{kg minute}^{-1}$

The final part of the student's answer is rather long-winded and hasn't really answered the question. The student is correct about the starting temperature being higher than the working temperature. However, he hasn't related this to the fact that at **a low temperature, the resistancc of the heating element is lower than at the working temperature, therefore the current will be higher.**

Don't make these mistakes ...

Don't think that examiner's can read your mind when you write down an answer. What may be obvious to you may well not be to other people. **Do write an explanation of what you are doing either with a few key words or else with mathematical symbols** such as:
∴ (therefore), ∵ (because), ⇒ (this implies) etc.

Do try to be precise when you are answering descriptive questions. Unless your handwriting is exceptionally large, the space or number of lines provided gives you an indication of the sort of quantity of text that the examiners are expecting. In the last part of the student's answer he had run out of space before reaching the important part of the answer.

Key points to remember

- Practical evidence for belief in the existence of gas molecules comes from **diffusion** and **Brownian motion**.
- The **ideal gas equation**: $pV = nRT$ is derived using the following assumptions about gas molecules:
  - Gases are composed of many randomly-moving molecules.
  - Intermolecular collisions are perfectly elastic – no kinetic energy is lost during a collision.
  - Molecules move in straight lines in between collisions.
  - The time taken for a collision is so small compared with the time between collisions, that you don't have to consider it.
- **Molecular velocities** are distributed according to the Maxwellian distribution and the mean square velocity of the gas molecules ∝ the temperature in kelvin ($\langle c^2 \rangle \propto T$) $pV = \frac{1}{3} Nm\langle c^2 \rangle$

  In these equations p = pressure in Pa, V is volume in m^3, N = number of molecules of gas, R = Universal molar gas constant (=8.3 $Jmol^{-1}K^{-1}$), T = temperature in K, m = mass of a molecule in kg, $\langle c^2 \rangle$ = mean square molecular velocity in ms^{-1}.
- Temperature in K = temperature in °C + 273

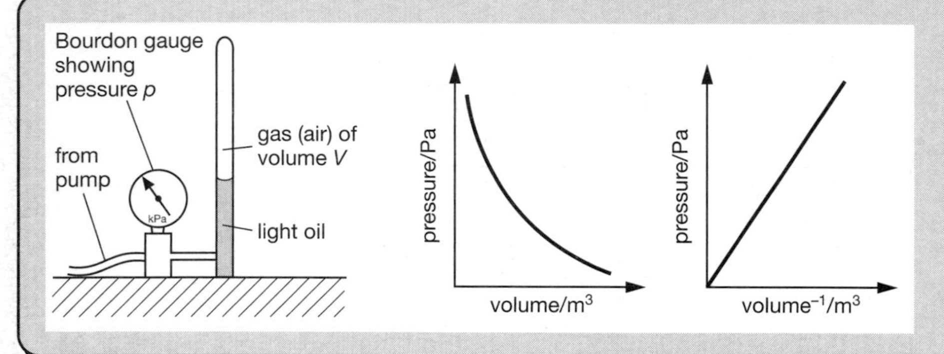

- This apparatus can be used to investigate Boyle's law $P \propto \frac{1}{V}$. A graph of p against V is a curve but p against $\frac{1}{V}$ gives a straight line origin graph

- When energy is supplied to a substance it can undergo a change in temperature or state.
- The change in temperature occurs because the internal energy of the substance increases due to increased random distribution of the energy between the molecules of the substance ($Q = mc\Delta\theta$).
- A change in state occurs because the energy supplied is required to break bonds between atoms (fusion) and expand against the surrounding pressure (vaporisation) ($Q = ml$).
- Here Q = energy transferred to substance by heating in J, m = mass of substance in kg, c = specific heat capacity in $Jkg^{-1}K^{-1}$, $\Delta\theta$ = temperature change in K and l = specific latent heat in J kg^{-1}.

The first law of thermodynamics is a restatement of the conservation of energy and is often written in the form:

$$\Delta Q = \Delta U + \Delta W$$

ΔQ = energy supplied to the system by heating
ΔU = increase in internal energy of the system
ΔW = work done by the system on the surroundings

A **heat engine** (such as a car engine) causes work to be done on the surroundings when energy passes from an object at high temperature (T_1) and it passes some energy to a second object at low temperature (T_2) by heating.

A **heat pump** (such as a refrigerator) is the reverse of a heat engine. Here work is done on the system to allow energy to pass from a system at low temperature to the surroundings at higher temperature.

The maximum thermal efficiency of a heat engine is given by

$$\eta = \frac{T_1 - T_2}{T_1}$$

Questions to try

Examiner's hints for question 1

In the first part, one of the variable should be obvious to you but you may need to think a little for the second ("chemistry" might give you a clue!)

You will need to plot at least two points in order to ensure that your curve is in the right region.

Q1

The relationship pV = constant applies to a sample of a gas provided that two other variables are constant. Name the two other variables.

....

....

[2 marks]

The graph below shows the variation of pressure p with volume V for an ideal gas at a temperature of 300 K.

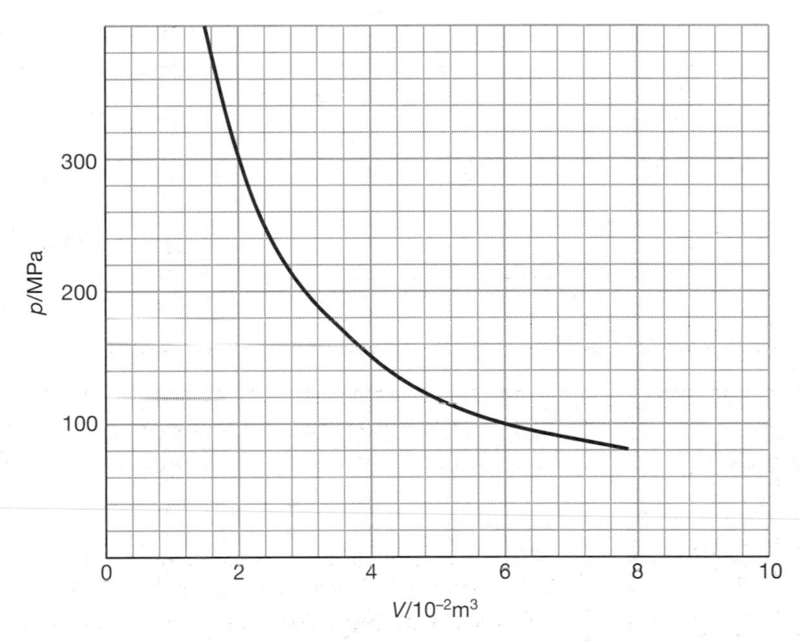

Add a second line to the graph showing the relationship between pressure and volume for the same sample of gas at a temperature of 400 K.

[2 marks]

[Total 4 marks]

Examiner's hints for question 2

Part (b): You need to consider the kinetic energy of one atom and then multiply this by the number of helium atoms in 1.0 g.

Part (c): This is really only another way of asking you what the relationship is between the mean kinetic energy of the gas atoms and the kelvin temperature.

Helium is a monatomic gas for which all the internal energy of the molecules may be considered to be translational kinetic energy.

molar mass of helium = 4.0×10^{-3} kg
the Boltzmann constant = 1.38×10^{-23} J K^{-1}
the Avogadro constant = 6.02×10^{23} mol^{-1}

(a) Calculate the kinetic energy of a tennis ball of mass 60 g travelling at 50 m s^{-1}.

....

....

[1 mark]

(b) Calculate the internal energy of 1.0 g of helium gas at a temperature of 48 K.

....

....

....

[3 marks]

(c) At what temperature would the internal energy of 1.0 g of helium gas be equal to the kinetic energy of the ball in part **(a)**.

....

....

[1 mark]

The answers to these questions are on pages 92 and 93. [Total 5 marks]

6 Answers to Questions to try

Notes: A tick against a part of an answer indicates where a mark would be awarded.

Chapter 1 Motion

Q1 How to score full marks

(a) Acceleration = change in velocity/time.
This is given by the gradient of the graph ✓
$= 20\ ms^{-1}/5\ s = 4\ ms^{-2}$ ✓

(b) Distance travelled = average speed/time
(= area under line) ✓
$= {}^1/_2 \times$ base $\times$ height
$= {}^1/_2 \times 4\ s \times 16\ ms^{-1}$
$= 32\ m$ ✓

Examiner's comment
Often with this sort of grid you would be expected to work out the area of one square and multiply by the number of squares. In this case the key word 'estimate' is missing and so you are expected to perform a calculation and to get an accurate answer.

Q2 How to score full marks

(a) 2.1s (number and unit) ✓

Examiner's comment
Two small squares represent 0.1 s and the ball changes direction every time it hits the ground so the third bounce is at the time corresponding to the third vertical line.

(b) The straight lines show that ✓ the gradient (acceleration) is constant ✓ and so the acceleration (due to gravity) is constant. ✓

(c) Distance is the area under the graph ✓ so the height will be area from the instant of dropping until it first hits the ground = 0.5 × 0.5 s × 5.0 m ✓
= 1.25 m

Examiner's comment
Alternatively you could use the equations of motion to calculate the distance. Note that it is the working that is important in this answer (not the actual answer).

(d)

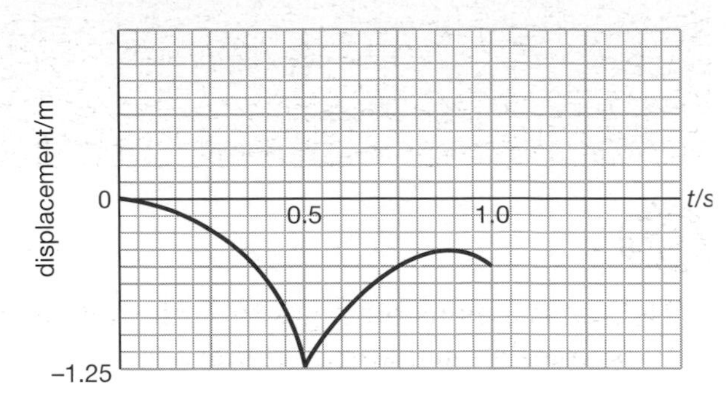

displacement scale as shown above ✓
correct curve shape (up to 0.5 s) ✓
Correct shape from 0.5 s (with peak below 0) ✓

Examiner's comment
This is a relatively difficult part of the question and it is likely that you would pick up marks for less that perfect graphs here. For example, if you had drawn positive displacement, you would only lose a mark (with everything else correct). Be aware that a mark scheme only offers one route to gaining full marks. Examiners will accept variations – providing the physics is not wrong or irrelevant.

(e) displacement = −1.25 m ✓

Examiner's comment
Displacement is a vector and it must have direction – downwards being negative (as with the velocity).

Chapter 2 Vectors and Projectiles

Q1 How to score full marks ✓

(a) Scalar quantities have magnitude alone whilst vector quantities have both magnitude and direction. ✓

(b) **(i)** 3.0 minutes = 0.05 hour ✓
distance = average speed × time
$100\ km\ h^{-1} \times 0.05\ h = 5.0\ km$ ✓

(ii) In this time the car has reached the side of the track opposite the starting point.

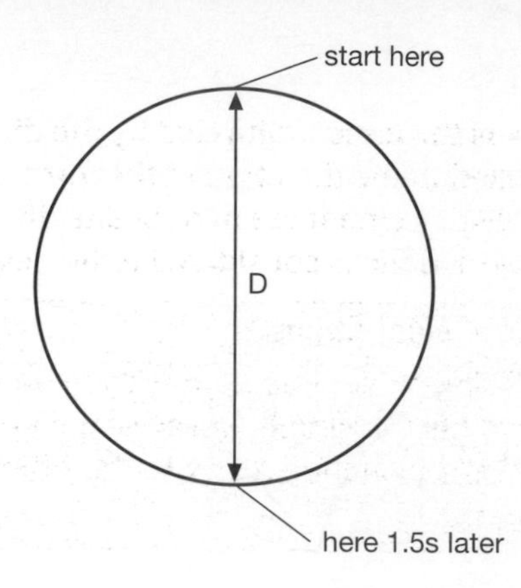

circumference = 5 km = πD ✓

D = $\frac{5.0 \text{ km}}{\pi}$

= 1.6 km ✓

Examiner's comment

Since the question asks for the magnitude of displacement, there is no need to give the direction; the worst data is given to 2 significant figures (sf) so your final answer should be no more accurate (examiners would usually accept 3 sf here).

Q2 How to score full marks

(a)

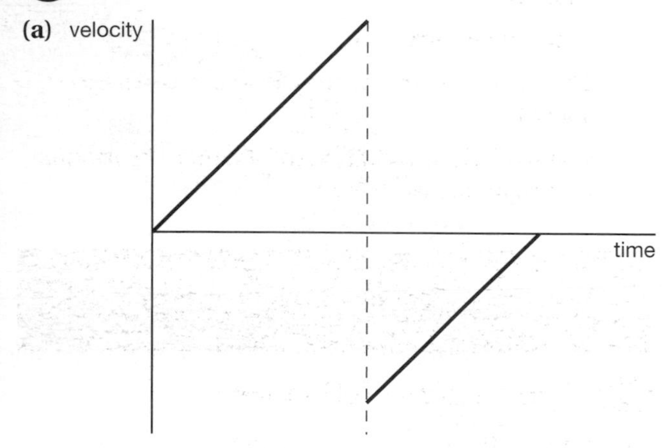

(b)

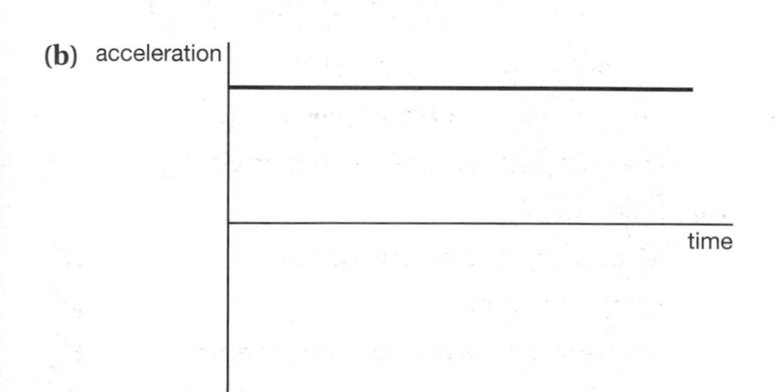

(a) straight line positive gradient ✓

velocity changes to negative value on bounce ✓

negative value smaller ✓

gradients the same ✓

(b) constant value for acceleration of gravity. ✓

(c) Since the acceleration of gravity close to the Earth does not change (in size or direction) (i) and (ii) will be vertically down. ✓ ✓

The momentum follows the direction of the total velocity and so will be along the tangent at P and Q. ✓✓

(d) horizontal component of the velocity = 15 ms^{-1} × cos 50°. ✓

= 9.64 ms^{-1} ✓

momentum = 0.15 kg × 9.64 ms^{-1} ✓

= 1.45 kg ms^{-1} or Ns ✓ (in the horizontal direction)

Chapter 3 Forces, Moments and Equilibrium

Q1 How to score full marks

(a) Total mass = 90 kg

$w = mg$ ✓ = 90 kg × 9.8 Nkg^{-1}

= 882 N ✓

(b) each vertical component = force × cos θ ✓

The sum of the vertical components = 200 N cos 30° + 550 N cos 14°

= 173 N + 534 N

=707 N (approximately 700 N). ✓

(c) There is no resultant force (or moment) acting on the climber. ✓

Examiner's comment

With a single mark, the examiners are likely to accept either force or moment here. If you are in any doubt, and since the answers are not contradictory, include both options.

(d) The vertical forces must add up to zero because the climber is in equilibrium.

Weight = total upward forces ✓

882 N = 707 N + F sin 37°

F = 175 N/ sin 37°

= 292 N ✓

(e) For the free foot to accelerate upwards it must experience an upward force. The other foot (or hand grips) would need to provide this and so the force on the foot would increase. ✓ **The free foot at this time would no longer be in equilibrium.** ✓

Q2 How to score full marks

(a) (i) moment of force about point = force (F) × perpendicular distance from the direction of the force to the point(x). ✓

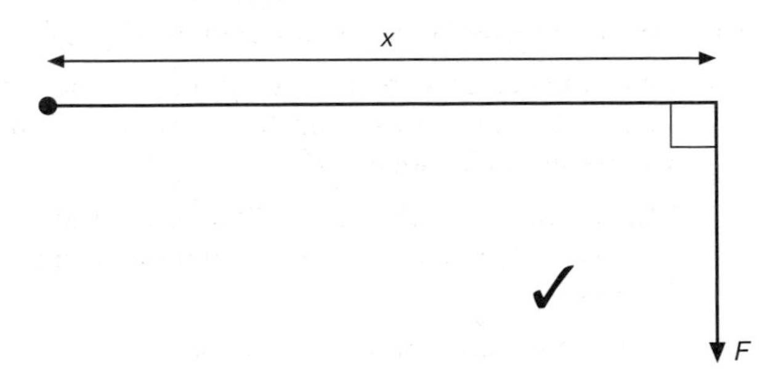

A couple is a pair of forces which are equal in magnitude and in opposite directions.

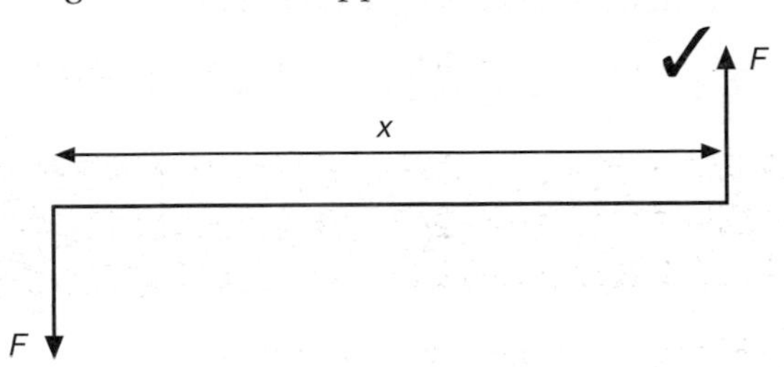

Torque (another name for *moment*) = the product of the one force × the perpendicular distance between them. ✓

(b) (i) moment = 280 N × 6 m ✓

= 1680 Nm ✓

(ii) Sum of clockwise moments = sum of anticlockwise moments ✓

1680 Nm = T × 4m sin 35° ✓

T = 732 N ✓

Chapter 4 Momentum, Work, Energy and Power

Q1 How to score full marks

(a) For a varying force the work done would be found by plotting a graph of the force against distance. ✓ The work done is the area under the curve. ✓

> **Examiner's comment**
> For mathematicians! It is not sensible to mention integration of the force function since this is beyond the requirements of AS physics.

(b) (i) Work is the force multiplied by the distance travelled in the direction of the force. If force and displacement are in opposite directions the work done is considered to be negative. ✓

> **Examiner's comment**
> Work being scalar has no direction so the negative here is simply a way of indicating the rise or fall of kinetic energy of the object in question.

(ii) The kinetic energy falls. ✓

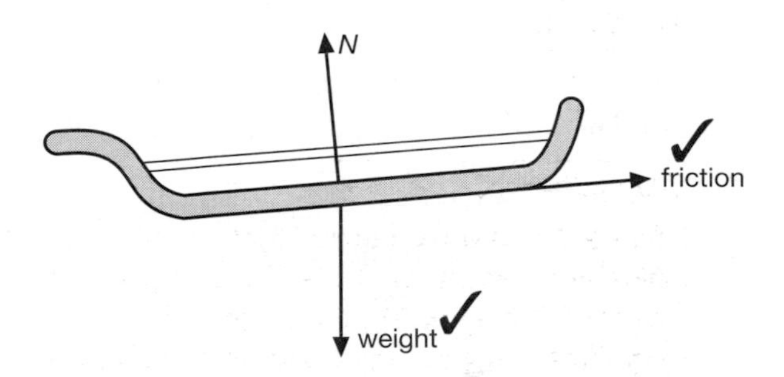

weight is produced by the gravitational pull of the Earth on the sledge

The snow rubbing against the sledge produces friction

(one cause correct) ✓

The resultant force is zero ✓ (since constant speed).

The work done by N is zero ✓ (since the motion is at right angles to N).

Chapter 5 Mechanical Properties of Matter

Q1 How to score full marks

(a) tensile stress = $\frac{\text{tension}}{\text{cross-sectional area}}$ ✓

tensile strain = $\frac{\text{extension}}{\text{original length}}$ ✓

Use of tensile (or tension) and original ✓

(b) Mention of:

Measuring the diameter of wire ✓

using a micrometer ✓

at several places or repeat and average ✓

Measuring the (original) length of wire ✓

using metre rule(s) or tape measure

(maximum of 4 marks)

(c) stress

✓ (plastic region)

✓ (linear region)

strain

For a stress/strain graph, the linear region gains one mark and the correctly curved plastic region the second mark.

Chapter 6 Direct Current Electricity

Q1 How to score full marks

(a) **(i)** charge = current × time ✓

(ii) coulomb (C) ✓

(iii) work done per coulomb ✓ in converting electrical energy into some other form of energy; ✓

volt = joule per coulomb ✓

(b) **(i)**

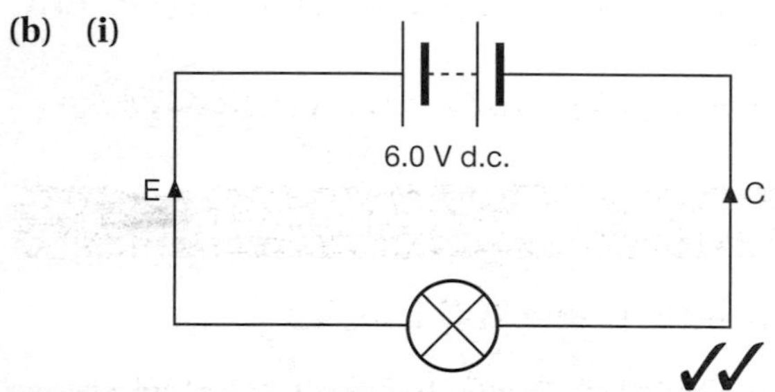

✓✓

(ii) W = QV ✓

= 90 ✓ (J) – unit already given

Q2 How to score full marks

(a) $R = \frac{\rho l}{A}$ ✓

$R = \frac{4.0 \times 10^{-3} \times 20 \times 10^{-3}}{\pi \times (0.70 \times 10^{-3})^2}$ ✓

(b)

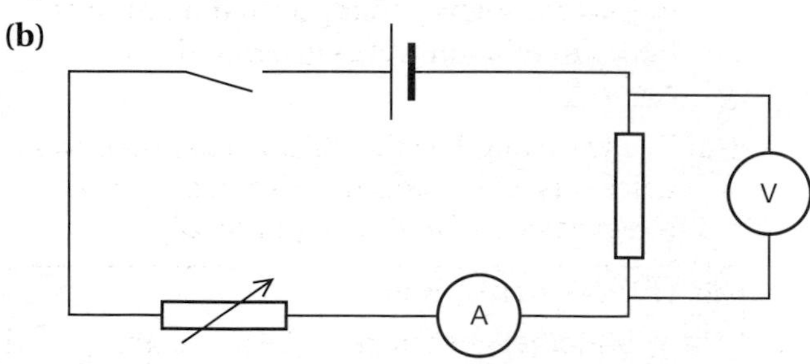

complete circuit with all components ✓

ammeter and voltmeter correctly connected ✓

The lead is the component here.

Adjust the variable resistor to measure a series (say 5 sets) of readings of current and the corresponding voltage. ✓

Plot a graph of voltage (y-axis) against current (x-axis). ✓

Calculate the gradient = resistance. ✓

Measure length of lead with ruler (l). ✓

Measure the diameter with a micrometer screw gauge and calculate the area from $A = \frac{\pi d^2}{4}$ ✓

Repeat and average all readings. ✓

> **Examiner's comment**
>
> Other variations of this method are quite feasible, e.g. use of an ohmmeter to measure the resistance directly. Examiners would give full credit for equivalent correct answers.

(c) $P = I^2R = (0.25\ \text{A})^2 \times 52\ \Omega$ ✓

$= 3.25\ \Omega$ ✓

> **Examiner's comment**
>
> It is not absolutely necessary to include units within calculations but it is an excellent habit to acquire since you can practise using correct units without even thinking about it!

Chapter 7 Voltages and Internal Resistance

Q1 How to score full marks

(a) *Potential divider*

advantage: allows the full voltage from $0 - V_s$ to be placed across the bulb ✓

disadvantage: not all of the current passes through the filament of the bulb ✓

rheostat

advantage: simple circuit to connect ✓

disadvantage: current through filament of bulb can never be zero (unless battery flat). ✓

(b) **(i)** $I = V/R = 2\ \text{V}/8\ \Omega$

$= 0.25\ \text{A}$ ✓

(ii) $P = IV$

$0.5\ \text{W} = \text{I} \times 2\ \text{V}$

I = 0.25 A (through bulb filament)

total current through AX = $I_{\text{bulb}} + I_{\text{XB}}$

= 0.5 A ✓

The last mark is well earned!

Q2 How to score full marks

(a) **(i)** The e.m.f. is the work done per coulomb of charge ✓

in converting energy from chemical to electrical internally ✓

(ii) When a current flows there will be a potential difference developed across the internal resistance – the *lost volts.* ✓

(b) $E = I(R + r)$

$12\text{ V} = 11.5\text{ V} + Ir$ ✓

$Ir = 0.5\text{ V}$

$I = \dfrac{11.5\text{ V}}{10\ \Omega} = 1.15\text{ A}$ ✓

$r = \dfrac{0.5\text{ V}}{1.15\text{ A}} = 0.43\ \Omega$ ✓

Chapter 8 Electrical Circuits

Q1 How to score full marks

(a) **Circuit P:**

$1/R = 1/R_1 + 1/R_2 + \ldots$

$= 1/24\ \Omega + 1/24\ \Omega + \ldots$ ✓

$R = \dfrac{24\ \Omega}{8} = 3\ \Omega$ ✓

$I = \dfrac{12\text{ V}}{3\ \Omega}$

$= 4.0\text{ A}$ ✓

Circuit S:

$R = 8 \times 0.5\ \Omega$ ✓

$I = \dfrac{12\text{ V}}{4\ \Omega}$

$= 3.0\text{ A}$ ✓

(b) **Circuit P:**

new $R = \dfrac{24\ \Omega}{6} = 4\ \Omega$

new $I = \dfrac{12\text{ V}}{4\ \Omega} = 3.0\text{ A}$ ✓

Circuit S:

circuit is now open so current = 0 ✓

(c) 12 V has fallen to 6 V thus I becomes $I/_2$ ✓

(P=) $I^2R \wedge \dfrac{I^2R}{4}$ power has become

$^1/_4$ of original power ✓

Answer to Kirchhoff's law example from page 44

substituting from **(i)** into **(ii)** gives:

$1.5\text{ V} = I_1 \times 2\ \Omega + (I_1 + I_2) \times 10\ \Omega$

therefore:

$1.5\text{ V} = I_1 \times 12\ \Omega + I_2 \times 10\ \Omega$............**(iv)**

substituting from **(i)** into **(iii)** gives:

$1.3\text{ V} = I_2 \times 3\ \Omega + (I_1 + I_2) \times 10\ \Omega$

therefore:

$1.3\text{ V} = I_2 \times 13\ \Omega + I_1 \times 10\ \Omega$............**(v)**

the I_1 term in **(iv)** is × 12 Ω and in **(v)** is × 10 Ω; so if we multiply **(iv)** by 5 we get:

$7.5\text{ V} = I_1 \times 60\ \Omega + I_2 \times 50\ \Omega$..............**(vi)**

and if we multiply **(v)** by 6 we get:

$7.8\text{ V} = I_2 \times 78\text{ W} + I_1 \text{ x } 60\ \Omega$............**(vii)**

subtracting **(vi)** from **(vii)** gives:

$0.3\text{ V} = I_2 \times 28\ \Omega$

$I_2 = 0.3\text{ V}/28\ \Omega = 0.011\text{ A}$

Substituting for I_2 into equation **(iv)** gives

$1.5\text{ V} -= I_1 \times 12\ \Omega + (0.011\text{ A}) \times 10\ \Omega$

so $I_1 = (1.5\text{ V} - 0.11\text{ V})/12\ \Omega = 0.116\text{ A}$

going back to equation **(i)** gives I_3 as 0.127 A (= 0.116 A + 0.011 A)

and the p.d. across resistor = 0.127 A × 10 Ω = 1.27 V

Chapter 9 Wave Properties

Q1 How to score full marks

(a) The particles are made to vibrate ✓ at the same frequency as the source. ✓ As particles get further away from the source the phase lags by an increasing amount. ✓ If the system is damped the amplitude decreases with distance. ✓

(maximum of 4 marks)

(i) The amplitude is the maximum displacement of the wave ✓ from the rest position.

(ii) The frequency of the wave is the number crests or troughs passing a fixed point in 1s ✓ or the number of waves generated per second.

(iii) The wavelength is the distance between two adjacent ✓ crests ✓ or the distance between two particles vibrating in phase. ✓✓

> **Examiner's comments**
>
> Many statements equivalent to these are correct – check yours against your notes or textbook.

(c) $v = f\lambda$ ✓
$v = 5.0\text{Hz} \times 3.0\,\text{m} = 15\,\text{ms}^{-1}$ ✓

Q2 How to score full marks

(a) Waves are transmitted as a result of total internal reflection ✓ (refraction but not reflection alone).

(b) The signal needs to be converted into an appropriate form using a transducer ✓

one from:
light signal down an optical fibre
electrical signal down a conducting cable
electromagnetic waves via a satellite ✓

(c) **one** from:
advantage of method in figure:
no specialised satellite needed therefore avoids cost of building and placing in orbit ✓

OR

disadvantage:
ionosphere is of varying composition therefore weather effects quality of signal ✓

> **Examiner's comments**
> There are many correct answers to this part and examiners would accept any appropriate answer provided it relates to the two methods of transmission mentioned in the question and in your answer.

Chapter 10 Waves and Sound

Q1 How to score full marks

(a) **(i)** The wave is a standing wave. ✓

(ii) The wave is transverse. ✓

(b) When the bow is dragged across the string it causes the string to vibrate. ✓ This sends a wave towards each fixed end. ✓ The waves reflect at each end. ✓ As the reflected waves meet they show superposition (or interfere) ✓ (forming a resultant wave of displacement equal to the vector sum of the individual displacements of the waves).

> **Examiner's comments**
> These are the key points but they could be phrased differently and still gain the marks.

(c) **(i)** Distance between nodes (= 1.25 m/5) = 0.25 m ✓

(ii) Wavelength (= twice distance between nodes) = 0.50 m ✓

(d) The wave is transverse and so it can be polarised. ✓ It may be difficult to polarise such a wave on a musical instrument since the strings are fixed and polarisers (slits for example) would be difficult to install. ✓ It is unlikely that the wave would be polarised in the first place since the bow would vibrate the string in several planes. ✓

> **Examiner's comments**
> With such an open-ended question, examiners would probably be flexible in their marking of this – as long as the two supporting points make sense and are relevant.

(e) In order to calculate the speed of the wave on the string you would need to know the frequency and the wavelength. ✓

To measure the frequency of the wave use a microphone connected to an oscilloscope. ✓

By adjusting the time base such that the shown wave pattern appears, you can calculate the period of the wave (= time corresponding to the distance between two nodes on screen). ✓

Frequency = 1/period ✓

λ is calculated as in part **(c)**. ✓

The speed is calculated from $c = f\lambda$. ✓

> **Examiner's comments**
> This is another example of how six relevant and correct points can score all six marks available.

Chapter 11 Waves and Light

Q1 How to score full marks

(a) The waves are diffracted by the two slits. ✓

At a bright fringe there is constructive interference. ✓

The waves are in phase when they reach the double slit. ✓

The path difference between waves from the two slits must be λ. ✓

The waves are still in phase at P. ✓

(b) **(i)** $y = \frac{\lambda D}{d}$ (where y = fringe separation, λ the wavelength, D the distance from the double slit to the screen and d the separation of the slits)

when D increases, y increases (everything else staying constant) ✓

(ii) when d increases, y decreases (everything else staying constant) ✓

(iii) when λ decreases, y decreases (everything else staying constant) ✓

(c) mention of constant phase relationship between waves ✓✓
(if marks not scored, mentioning that the waves have a constant frequency may score a consolation mark)

Chapter 12 Photons, Energy Levels and Spectra

Q1 How to score full marks

a) (i) The line emission spectrum consists of a series of coloured lines ✓ on a dark background. ✓

(ii) The line absorption spectrum consists of a series of dark lines ✓ on a (continuous visible) spectrum. ✓

(b) The coloured lines in **(a)(i)** correspond to the position of the black lines in **(a)(ii)**. ✓

(c) (i) When atoms absorb energy, electrons are excited to higher energy levels. ✓ When the electrons fall to a lower energy level each emits a photon of energy ✓ (that is equal to the difference between the energy levels for that transition).

When white light passes through cool vapour the wavelengths corresponding to energy level transitions are absorbed ✓ meaning that when this light is re-radiated the intensity in the original direction is reduced. ✓

(ii)

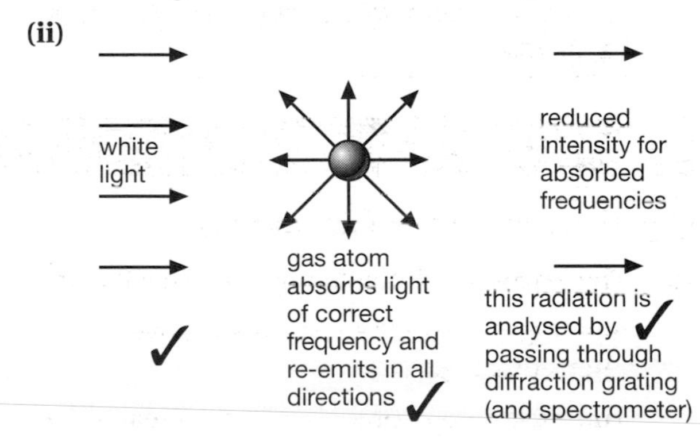

Examiner's comments

This answer could be written – but a simple diagram is better.

(d) (i) longest wavelength ⇒ least energy difference ✓

$\Delta E = (2.06 - 1.89)$ eV = 0.17 eV ✓

0.17 eV = $0.17 \times 1.6 \times 10^{-19}$ J = 2.72×10^{-20} J

$$\lambda = \frac{hc}{E} = \frac{6.6 \times 10^{-34}\,\text{Js} \times 3.0 \times 10^{8}\,\text{ms}^{-1}}{2.72 \times 10^{-20}\,\text{J}}$$ ✓

$= 7.28 \times 10^{-6}$ m ✓

(ii) There are six ✓ possible lines from these four energy levels.

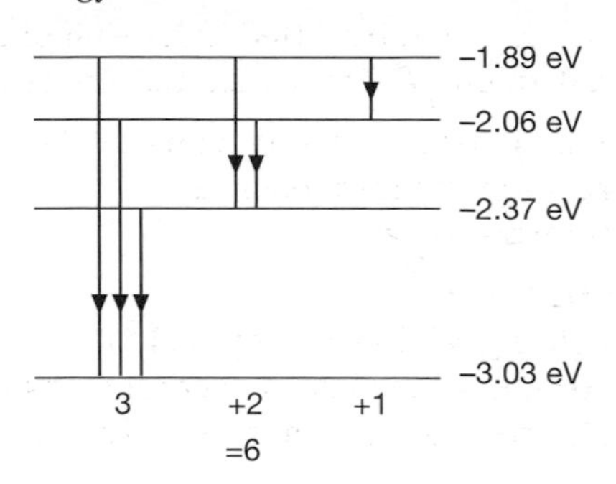

(e) (i) Mercury discharge lamps give out ultra violet light. ✓

(or $\lambda = \frac{c}{f} = 3.0 \times 10^{8}\,\text{ms}^{-1}\ /\ 2.5 \times 10^{15}\,\text{Hz}$

$= 1.2 \times 10^{-7}$ m ⇒ u.v.)

(ii) When atom absorbs the highest frequency that it would emit this is the ionisation energy.

ionisation energy = hf
$= 6.6 \times 10^{-34}\,\text{Js} \times 2.5 \times 10^{15}\,\text{Hz} = 1.65 \times 10^{-18}\,\text{J}$ ✓

$$\text{ionisation energy} = \frac{1.65 \times 10^{-18}\,\text{J}}{1.6 \times 10^{-19}\,\text{C}} = 10.3 \text{ eV}$$ ✓

Chapter 13 Radioactivity

Q1 How to score full marks

The corrected count rates will be:

Time, t / s	0	30	60	90	120
Corrected Count rate / counts per second	13.6	9.2	6.2	4.2	2.8

✓ for all values correct.

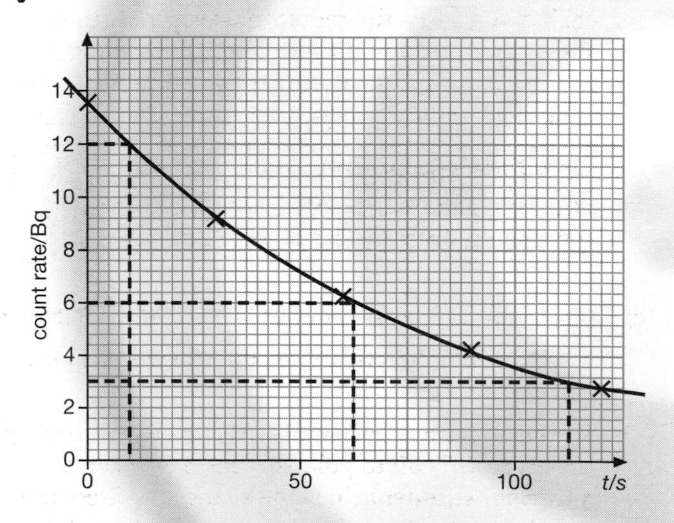

The corrected count rate is plotted accurately ✓ against time and a smooth curve drawn through the points ✓.

The half-life is clearly marked ✓ in twice (from 12 counts per second (cps) → 6 cps and from 6 cps → 3 cps) – this is good practice!

These values are clearly marked on the graph and the corresponding times found to be (62.5 – 10) s = 52.5s and (112.5 – 62.5) s = 50 s.

The mean half life is 51 s to the nearest second. ✓

Q2 How to score full marks

> **Examiner's comments**
> This sort of question is an ideal one in which to assess your quality of written communication so do check your English!

(a) In this case the answer is given as a list to make it easier for you to check your answer.

The nucleus is small compared with the atom.
The nucleus is extremely dense.
The nucleus is positively charged.
The electrons surround the nucleus.
The electrons are negatively charged.
The electrons are very small or much less massive than the nucleus.

(each point ✓ maximum 5)

(b) **(i)** When the α particle is close to the gold nucleus ✓ the positive charges on the nucleus **and** the α particle ✓ repel each other ✓ (hence significant angles of deflection).

(ii) Since the majority of the gold atom is empty space ✓ it is unlikely that many α particles pass close enough to a gold nucleus to allow significant deflection ✓.

Chapter 14 Particles

Q1 How to score full marks

(a) **(i)** There is one up quark ✓ and there are two down quarks ✓ in a neutron.

(ii) The neutron is neutral and so this arrangement of three quarks (as required by baryons) gives a charge of zero ✓ **and a baryon number of one.**

(b) **(i)** neutron → β + <u>proton</u> ✓ + <u>anti-neutrino</u> ✓

(ii) One of the down quarks has decayed into an up quark ✓ changing the charge from zero ($= -\frac{1}{3} + -\frac{1}{3} + \frac{2}{3}$) into +1 ✓ ($= -\frac{1}{3} + \frac{2}{3} + \frac{2}{3}$).
The baryon number remains 1.

Q2 How to score full marks

(a) Hadrons experience the strong (and the weak) nuclear force. ✓

Leptons experience the weak nuclear force. ✓

(b) Hadrons are (baryons such as) protons ✓ and neutrons and (mesons such as) pions and kaons.

Leptons are electrons, ✓ mus, taus and their anti-particles and neutrinos.

(only one correct example of each to score)

(c) Baryons are composed of 3 quarks. ✓

Mesons are composed of 2 quarks. ✓

(d) **(i)** A neutron is composed of an up quark and two down quarks. ✓

(ii) One of the down quarks decays into an up quark. ✓

(iii)

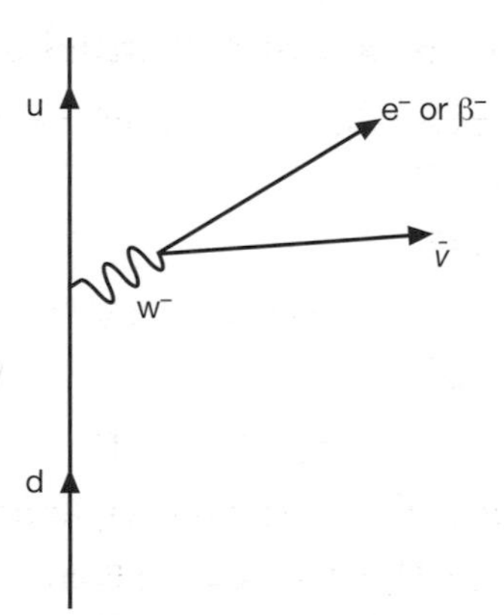

The diagram should show the emission of the W⁻ ✓ boson with the subsequent decay into the β particle **and** the anti-neutrino. ✓

Chapter 15

Q1 How to score full marks

> **Examiner's comments**
> The ideal gas equation is $pV = nRT$ and so the product pV will be constant if n and T are constant (since R is the Universal Molar Gas Constant)

So temperature ✓ and amount of gas ✓ (number of moles or mass of gas) must also be constant.

Since the gas in the second part obeys the ideal gas equation then any value of p on the curve multiplied by the corresponding value of V will be a constant × T. For the curve given (ignoring powers) the value is 600 therefore when T = 300K the constant will be 2.
At 400 K, since it is the same gas, the product will be 2 × 400 K = 800. The curve must go through the points (2 × 10^{-2} m^3, 400 MPa) etc. as shown on the next page.

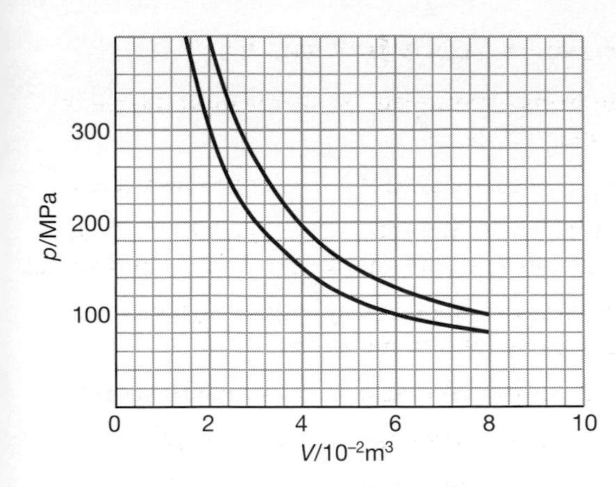

for the curve drawn above the first curve ✓

for the curve passing through at least two correct points ✓

Q2 How to score full marks

(a) The mean kinetic energy of the tennis ball is given by:

$E_k = \frac{1}{2} mv^2$

$= 0.5 \times 6 \times 10^{-2}\,\text{kg} \times (50\,\text{ms}^{-1})^2$

$= 75\,\text{J}$ ✓

(b) internal energy of 1.0 g of He

$= \frac{3}{2} nRT$ ✓ (where n is the number of moles of gas)

($R = k \times N_A$ where k = the Boltzmann constant and N_A = the Avogadro constant).

$= \frac{3}{2} \times \frac{1\,\text{g}}{4\,\text{g}} \times k \times N_A \times 48\,\text{K}$

$= \frac{3}{2} \times 0.25\,\text{mol} \times 1.38 \times 10^{-23}\,\text{JK}^{-1} \times 6.02 \times 10^{23}\,\text{mol}^{-1}$ ✓

$= 150\,\text{J}$ ✓

There are other ways of obtaining the same answer as this that are equally valid, e.g. using: internal energy $= \frac{3}{2} NkT$ (where N = number of atoms of helium)

(c) Since internal energy $\propto$ temperature (in K)

in order to halve the internal energy the temperature must also halve

$\therefore T = 24\text{K}$ ✓

Notes

Notes

ERRATUM

Some errors have unfortunately occurred in the setting of this book. Please note that the following equations should read:

p.29 Part (c) (ii)
$E = \sigma/\varepsilon$

p.34

- **Current** $I = \dfrac{\Delta Q}{\Delta t}$
- **Potential difference** (p.d.) $V = \dfrac{W}{Q}$
- **Resistivity** $\rho = \dfrac{RA}{l}$

p.44
$R = \dfrac{R_1 \times R_2}{R_1 + R_2}$

p.88 Q2 (a) $R = \dfrac{\rho l}{A}$